PRAISE FOR *SUP*… *ANALYTICS AN*…

This book gives comprehensive insights into business analytics and supply chain modelling, while inspiring views and vision on future trends of supply chain analytics for both students and professionals engaged in the analysis of complex and rapidly changing supply chains. Nicoleta Tipi provides the reader with a practical view on how to utilize advances in supply chain analytics and modelling in order to tackle modern complex supply chain activities which is highly useful for companies to have a competitive edge.
Sara Elgazzar, Dean of the College of International Transport and Logistics, Arab Academy for Science and Technology and Maritime Transport

This terrific book is essential reading for both new and experienced professionals. It's practical and full of invaluable information. I wish I had had this information when I was doing my PhD research.
Georgakoudis Elias, CEO, Multi Pack

Supply chain complexity is increasing. As such, this is a well-timed book that captures the key issues associated with navigating, analysing and evaluating data across the supply chain. It contains valuable insights that will be relevant to both practitioners and academics involved in supply chain management.
Helen Rogers, Professor of International Management, Nuremberg Institute of Technology

In this complex and rapidly changing time in supply chain management, in part due to the 4th Industrial Revolution, Nicoleta Tipi has provided logisticians and supply chain directors with insightful and comprehensive knowledge about measuring supply chain performance and business analytics. Put this book on your required-reading list. You will learn data analytics, data visualization, and other models to meet future challenges. This book is a helpful guide to the supply chain modelling field, providing both coverage and detail.
Khaled El Sakty, Dean of Faculty, Arab Academy for Science and Technology and Maritime Transport

Supply Chain Analytics and Modelling was much needed. This book will be of interest for students as well as practitioners who want to understand and improve their knowledge and capability on supply chain analytics and modelling.
Ozlem Bak, Senior Lecturer in Operations Management, Brunel University London

Supply Chain Analytics and Modelling provides a solid foundation in business analytics. It connects data analysis and how it can be further integrated with supply chain modelling providing potential to improve the decision-making process within a supply chain context. Suitable for undergraduate as well as postgraduate level studies in supply chain modelling and business analysis.
Ahmed Tarek El-Said, Senior Teaching Fellow, Warwick Manufacturing Group, University of Warwick

Provides a holistic view and deep insights about supply chain and its related concepts, bringing together valuable perspectives on its theoretical and practical application. Furthermore, the book also provides a very thoughtful inside on how to analyse data in the supply chain to obtain valuable business models and to develop proper strategies. The book is a must-read for academics and practitioners dealing with the problems and challenges of data analysis in the supply chain.
Dan-Cristian Dabija, Professor of Marketing, Babes-Bolyai University Cluj-Napoca

Supply Chain Analytics and Modelling is a great insight into big data and analytics modelling for global supply chains and helps the reader to understand the challenges faced by individuals and companies in today's ever-evolving data-centric environment. The book provides excellent advice on the importance of data to build effective and resilient supply chains by showing how to use the data for KPIs and analysis of your supply chain, which is a topic many teams struggle to know where to start with. With several excellent examples of how to obtain data, assess it and present it back, this is an excellent guide. A book well worth reading but also a reference guide, from an author with years of experience.
Charlotte Waddon, Senior Business Analyst Global Supply Chain

Supply Chain Analytics and Modelling

Quantitative tools and applications

Nicoleta Tipi

First published in Great Britain and the United States in 2021 by Kogan Page Limited

2nd Floor, 45 Gee Street
London
EC1V 3RS
United Kingdom

122 W 27th St, 10th Floor
New York, NY 10001
USA

4737/23 Ansari Road
Daryaganj
New Delhi 110002
India

www.koganpage.com

Kogan Page books are printed on paper from sustainable forests.

ISBNs

Hardback 9780749498627
Paperback 9780749498603
Ebook 9780749498610

British Library Cataloguing-in-Publication Data

A CIP record for this book is available from the British Library.

Library of Congress Control Number

2021003077

Typeset by Integra Software Services, Pondicherry
Print production managed by Jellyfish
Printed and bound by CPI Group (UK) Ltd, Croydon CR0 4YY

CONTENTS

LIST OF FIGURES AND TABLES

Figures

Tables

PREFACE

The concepts presented in this book are the author's own interpretation, formed from a number of years working with these issues from a practical, theoretical and professional perspective. A number of examples are used, and some of these are hypothetical in nature; however, they are intended to form the base of the explanation in each case. Specific concepts on business analytics, supply chain modelling, supply chain performance measures and big data in the supply chain will be introduced where they will be supported by definitions and relevant references at the end of each chapter. Additional sources will also be provided that form the base of the reading for a particular topic. The aim of this book is to introduce the reader to old and new concepts in business analytics, to challenge their interpretation and application, and to discuss, explain and promote new avenues regarding the salient characteristics and features of analytics in the context of supply chain systems.

This book is an introduction to business analytics and modelling within the field of supply chain and could be used by:

- undergraduate and postgraduate students of business intelligence, business analytics, data analytics, modelling and simulation in the supply chain;
- undergraduate and postgraduate students of business, management, technical and engineering universities as additional reading;
- PhD researchers of business, management, information technology, information systems and engineering;
- researchers who specialize in analytics, modelling and simulation and are concerned with their applications in general;
- practitioners who are keen to understand, evaluate, implement and review some of the related techniques presented here.

The first four chapters form *Part One* of the book, which sets out the basic concepts that are required in modelling and analysing a range of different aspects present in a supply chain system.

Chapter 1 aims to introduce the meaning of business analytics and to take a critical view at how is this seen in the current business environment. Opportunities, challenges and limitations to business analytics are discussed.

The entire book focuses on business analytics and modelling in the supply chain, but this chapter covers the main concepts of business analytics with the view to introducing the remaining chapters that focus on the supply chain.

Chapter 2 starts by providing a definition of supply chain systems and networks. Following this, the content of the chapter will delve into understanding what is modelling, but more prominently what is modelling from the perspective of a complex supply chain point of view. Interlinked ideas of analytics and modelling for complex supply chain systems that form complicated network structures are challenged in this chapter. Examples of models and modelling characteristics are embedded in this discussion.

Chapter 3 reviews the importance of data in the supply chain. Different characteristics of data are presented in this case with the view to emphasize the importance of data and data behaviours when this is being incorporated in analysis. The way in which data is structured and managed in the supply chain will be challenged in this chapter.

Following these three chapters on analytics, modelling and data in the supply chain, *Chapter 4* continues with a discussion on identifying and setting up performance measures that meet supply chain requirements. These measures aim to be set from the point of view of the data present in the chain to the challenges in using modelling and analytics to set up and analyse performance measurement systems. Characteristics of design and implementation of performance measures and supply chain performance measurement systems will be explained.

Part Two looks at establishing the key important characteristics associated with current models, techniques and approaches employed in modelling and analysis. Therefore, *Chapter 5* takes a more detailed view on data and the elements required to visualize the results of analysis, or indications from measurement systems. Different visualization examples are considered in this chapter, with examples presented from data being visualized using Excel and Minitab.

Chapter 6 looks at some of the most-used business analytics models under the descriptive and predictive analytics categories, and explains how these could be used in practice. Individual examples are detailed in this chapter that again are represented with the use of Excel and Minitab.

Complex models on supply chain analytics are then taken further in *Chapter 7*, where they are presented from an optimization and heuristics modelling perspective. Examples of linear programming considering the assignment, transportation, covering and the transshipment concept are detailed in this section. All these are discussed from a supply chain context

Figure 0.1 Business analytics and modelling in the supply chain

Part One

Business analytics

Chapter 1

Data in the supply chain

Chapter 3

Modelling in the supply chain

Chapter 2

Supply chain performance measurement systems

Chapter 4

Supply chain system

Part Two

Supply chain analytics (prescriptive models)

Chapter 7

Business analytics models (descriptive and predictive models)

Chapter 6

Visualization in the supply chain

Chapter 5

Part Three

Future research agenda for supply chain analytics

Chapter 8

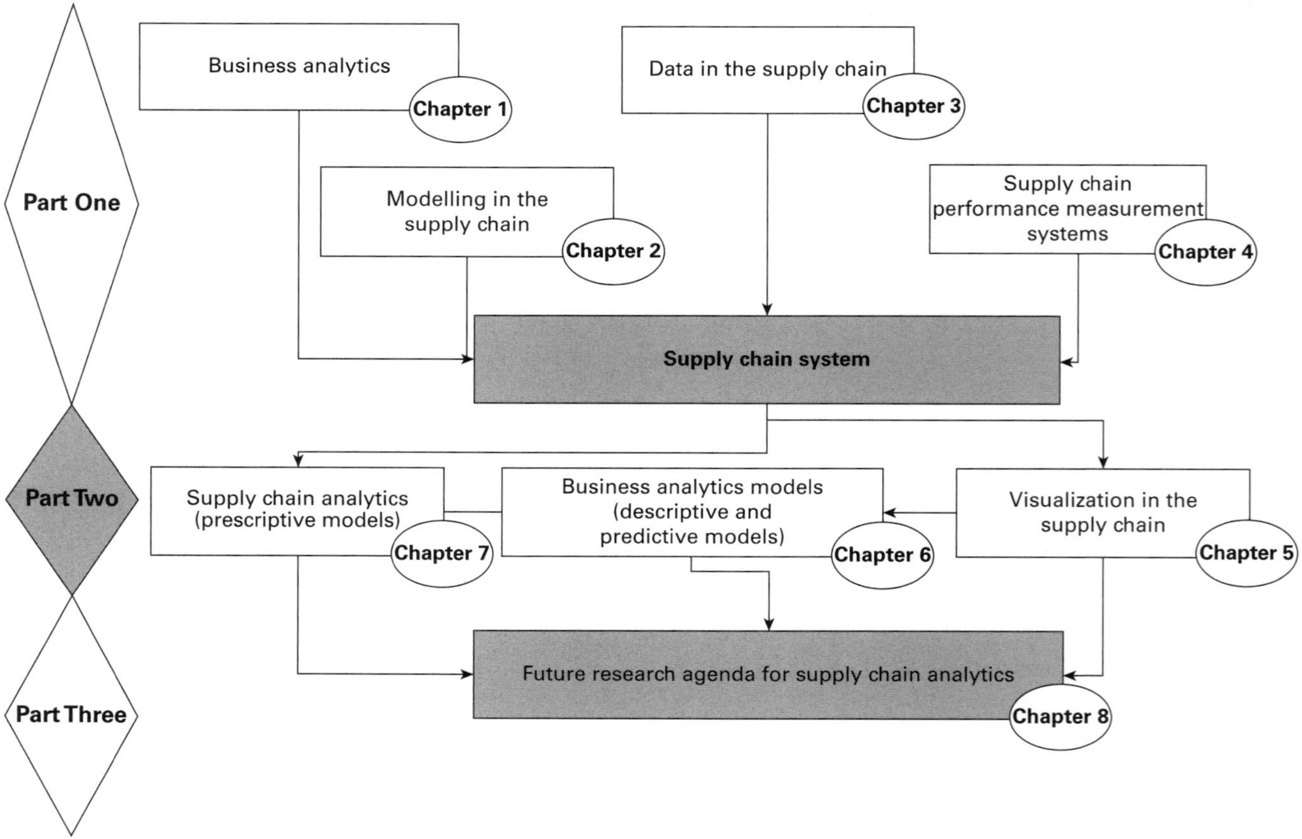

point of view. Heuristics models dealing with facility location and vehicle routing modelling are incorporated in this chapter. All these models are grouped under the prescriptive analytics section.

Following the details presented in the last three chapters, *Part Three* outlines new directions in the area of modelling and analytics. *Chapter 8* presents and challenges the future of business analytics and modelling. This is extracted from current systematic literature reviews conducted in the field of supply chain analytics and modelling.

PART ONE
An introduction to business analytics and modelling in the supply chain

Defining business analytics

01

LEARNING OBJECTIVES

- Enquire about and reflect on questions related to business analytics.
- Provide an understanding of what business analytics is and observe the different forms business analytics takes.
- Identify applications of business analytics.
- Evaluate key challenges in using business analytics.

Introduction

Analysing data, constructing different models that evaluate different aspects of already existing data, developing models that will have the power to predict behaviours and give further understanding into the labyrinth of data all form part of analytics. Providing further understanding of what represents business analytics will allow the exploration of what is to be grouped under the notion of business analytics and where the boundary lies with management science, operations management and operational research. Commonality and differences exist under each umbrella; however, the intention is to look at these from a business point of view and later on from a supply chain point of view. This is to try to provide an understanding of what to do in particular situations, how to develop an appreciation of very complex sets of data, how to use models to enhance business understanding and how to identify data and models that would contribute to creating business value.

Table 1.1 Sample of business analytics questions

Questions	Relevant to
What is business analytics? What is advanced analytics? What is prescriptive analytics? What is the rationale/purpose for undertaking business analytics? (Holsapple et al, 2014)	Researchers, practitioners, organizations
What are the key aspects to be considered when adopting business analytics within an organization?	Organizations: IT managers, finance department, software developers, business analytics vendors
How should the business needs of implementing business analytics (*within an organization*) be evaluated?	Organizations: strategic and operations managers, project managers, data implementation managers, business analysts Researchers: in the area of operations management, operational research, business analytics and supply chain
What data is available to carry out analytics (*internal and external to an organization*)? What are the current performance measurement systems? What IT/ERP system is in place to collect, store, generate and manage data?	Organizations: IT managers, data warehouse managers, database specialists, business analysts
What data is required to carry out business analytics? What performance measurement system needs to be put in place to best reflect current business operations and predict future business needs?	Organizations: strategic and operations managers, project managers, data warehouse managers, business analysts Researchers: in the area of operations management, operational research, business analytics and supply chain
How can the results of analysis be verified, validated and tested?	Organizations: data warehouse managers, IT managers, business analysts Researchers: in the area of operations management, operational research, business analytics and supply chain

What needs to be considered to implement the outcomes of carrying out analysis within an organization? What IT resources are required for the successful implantation of business analytics? What skills are required to develop, implement, analyse and maintain business analytics processes?	Organizations: strategic and operations managers, project managers, data warehouse managers, IT managers, business analysts, finance managers, HR managers Researchers: in the area of operations management, operational research, business analytics and supply chain
What are the critical success factors for implementing business analytics? What leads to post-implementation business analytics success?	Organizations: strategic, tactical and operations managers Researchers: in the area of operational research, operational management, strategic management, behaviour science, logistics and supply chain
How can business analytics be used to promote innovation?	Organizations: strategic, tactical and operations managers, engineers Researchers: in the area of operational research, operational management, strategic management, behaviour science, logistics and supply chain

Business analytics questions identified in the literature

Questions	**Source**
How can business analytics technologies be assimilated for competitive advantage?	Wang et al (2019)
What do the results of the analytics indicate for the manager responsible on the ground? Is analytics validated? Does analytics comply with the values of the company? Should analytics be used independently or in conjunction with other related aspects?	Bag (2016), p 24

Companies hold large amounts of data; they also continuously generate new data, and data is generated externally about or in relation to a company that influences different aspects of their business. The question that would need to be considered in this context, is what is to be put in place for an organization to hold only the required data, to collect only the data that leads to continuous improvement, and to seek externally generated data that brings value to them.

In many cases, data exists, where mathematical models are developed to provide further understanding of the existing data. The models used in this case could have different forms, such as to enquire, to predict, to correlate, to evaluation, to simulate, and so on. Model are not only there to evaluate data, they are also there to test processes, to evaluate different alternatives of a given situation, to evaluate whether a particular approach to change is the most appropriate. In any situation, models will not bring any benefits if they are not using relevant, reliable, timely and accurate data.

Therefore, this chapter aims to provide an understanding of the concept of business analytics (BA) in general and to contextualize the notion of business analytics from an organization's point of view. This chapter will also challenge the perceived opportunities offered by business analytics in the current business environment and highlight the limitations.

A set of questions could be put forward regarding business analytics, and a sample of these that could be relevant to researchers, practitioners, analysts and organizations is listed in Table 1.1.

What is business analytics?

Analyses of data have been carried out for a number of years; however, the amount of data recently generated has opened new opportunities for academics and practitioners to explore new avenues with data-driven decision-making tools. A number of statistical analytical methods have been trialled so far and many will continue to be used in the future. However, the rapid changes in technology have opened access to a new battlefield, where some of the traditional methods may be limited in their ability to capture the complex nature of current business needs. Data has been analysed for a number of years; where business analytics is currently receiving renewed attention, however, is through the interest to acquire, implement and extract value from data. Analytics has been studied in different forms in business schools over a number of years such as statistic, econometrics, financial analysis, simulations, linear programming and optimizations, vehicle routing and others.

Business analytics makes use of technology to support the decision-making process. However, it is not a technology on its own. Business analytics is a combination of tools, approaches and procedures that will have a holistic view in collecting relevant details and information, presenting these in the most appropriate way and predicting what could happen, and it has the ability to suggest optimum solution for any identified situations (Al-Haddad et al, 2019; Barbosa et al, 2017).

At the same time, business situations have been modelled from a number of years, where the results of these models have proven beneficial to organizations implementing them. Business analytics is comprised from a number of models and techniques. A key ingredient in gaining benefits from a model is the use of relevant, reliable and accurate data. Without the use of real data, the model's results will be just an abstraction, relevant only for testing, experimentation and learning. To gain practical benefits, organizations are keen to use reliable, robust models that could accept real data and are ready to be implemented in a practical setting.

Model implementation is one concern of any organization. The application of models will generate a set of results and provide the opportunity for a number of different alternatives. Implementing the most appropriate answers from these analyses, and ensuring that value is achieved over a period of time from the implemented results, is what brings the real benefit to an organization.

Extracting value from data has been discussed for a number of years. In a paper from 2004, Neely and Jarrar discuss the performance planning value chain (PPVC), a tool from the Centre of Business Performance, Cranfield School of Management. Starting with defining a hypothesis for what the data should be analysed for, and moving through six different stages to ensure value is extracted from the data gathered, the tool is presented with the view that a number of action planning tools, decision-making, feedback systems, project managements and prioritization techniques are imperative.

A number of steps have been identified by Abai et al (2019) when organizations may consider implementing business analytics. To start with, the problem and the opportunities should be identified, followed by understanding the data, collecting the required data, transforming this data to support the decision-making process and analysing the data. The steps regarding the data management are discussed further in Chapter 3. Abai et al (2019) continue by indicating steps such as developing a model, evaluating a model, using the model and translating the output. They also mention measuring the impact of the model created and maintenance, but relate that these two elements receive less attention from researchers.

Holsapple et al (2014) bring forward three dimensions when discussing business analytics. These are the domain, orientation and technique. The domain refers to the subject area to which analytics is being applied. A business analytics domain may include environmental analytics, customer analytics, supply chain analytics, web analytics, talent analytics, financial analytics and others (Holsapple et al, 2014). The orientation domain discussed here refers to the 'direction of thought' (Holsapple et al, 2014). This may include descriptive, predictive or prescriptive analytics, with some examples as indicated in Table 1.2. The technique domain is regarded as the way in which analytics is being performed.

To be in a position to master these tools and techniques, analyse the results and convey a reliable implementation, organizations need to employ skilled professionals to carry out these activities. In the past, analytics and analytical skills may have been required in a dedicated department within an organization; however, nowadays analytics is present in every aspect of an organization. Stanton and Stanton (2020) highlight in their study that currently organizations struggle to find employees with the required skills to manipulate and master advanced analytics. They mention that there is a gap between the number of positions in the field of analytics that are waiting to be field and skilled professionals who have a good understanding of the application and use of advanced analytics and at the same time understand the business needs and organization requirements to create long-term benefits.

In general, as depicted in Figure 1.1, the main categories we see are *descriptive*, *predictive* and *prescriptive analytics*, with a number of different categories added for individual research enquiries.

Descriptive analytics

With descriptive analytics, we tend to look at what happened or what is currently happening and why it happened or is happening. Inquisitive or diagnostics analytics could be grouped under this category as well. Authors such as Wang et al (2018) highlight the question: Why is it happening? and group these techniques in diagnostic analytics, where Sivarajah et al (2017) name these analyses inquisitive analytics when the same question is being asked. Descriptive analytics applications are in the form of reports, dashboards analysis and performance indicators (Appelbaum et al, 2018) and tend to collate data already available within an organization's IT system (Sukhobokov, 2018).

Table 1.2 Categories of business analytics

Categories of business analytics	Authors (year)
Descriptive analytics with queries such as: What happened? What is happening now? Why is it happening? Why has it happened? **Predictive analytics** with questions such as: What if these trends continue? What will happen next? What is likely to happen? **Prescriptive analytics** with the approach of: What could be done based on what happened?	Wang et al (2016); Nguyen et al (2018)
Descriptive analytics including techniques such as mapping and visualization. **Predictive analytics** can comprise time series methods (for example moving average, exponential smoothing, autoregressive models), linear, non-linear and logistics regression models; data-mining techniques (cluster analysis, market basket analysis). **Prescriptive analytics** could include analytic hierarchy process (AHP) technique, game theory (eg auction design, contract design); mixed-integer linear programming (MILP), non-linear programming, network flow algorithms, stochastic dynamic programming.	Souza (2014)
Descriptive analytics on questions such as: What has happened? **Inquisitive analytics**: Why is it happening? Why has it happened? ('*Probing data to certify/reject business propositions*') **Predictive analytics**: What will happen? **Prescriptive analytics**: What recommendations could be given? **Pre-emptive analytics** ('*Having the capacity to take precautionary actions on events that may undesirably influence the organisational performance*')	Sivarajah et al (2017, p 266)

(continued)

Table 1.2 (Continued)

Categories of business analytics	Authors (year)
Descriptive analytics refers to ad hoc reporting and queries. **Diagnostic analytics** explores hidden factors of a bad outcome and considers 'what if' analysis. **Predictive analytics** works with statistical methods and machine learning for forecasting and aims to answer questions on what will happen. **Prescriptive analytics** incorporates quantitative optimization techniques and stochastic simulation.	Wang et al (2018)
Descriptive analytics (reports, dashboards, OLAP-analysis, descriptive models). **Predictive analytics** (predictive analytics, implicative analytics, graph analytics). **Prescriptive analytics** (selection, ranking, allocation, action planning). **Executive analytics** (search in state space, reinforcement learning, lifelong learning, imitation learning, systemic learning). **Reflexive analytics** (machine theory of mind, multi-agent system of artificial intelligence). Artificial intelligence of human level.	Sukhobokov (2018)
Descriptive analytics **Predictive analytics** (probabilistic models, machine learning/data mining models and statistical analyses models). **Prescriptive analytics** (probabilistic models, machine learning/data mining, mathematical programming models, evolutionary computation models, simulation models and logic-based models).	Lepenioti et al (2020)

Figure 1.1 Basic representation of business analytics characteristics

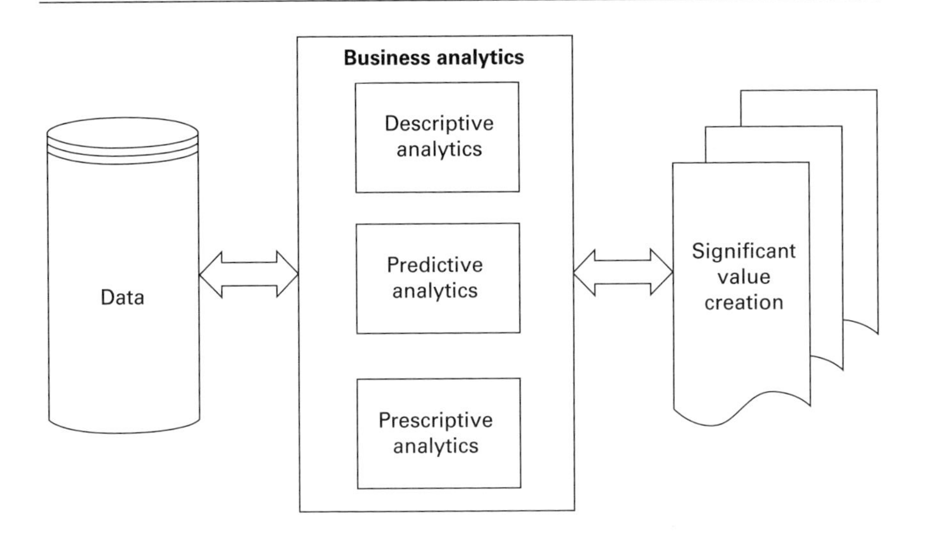

The data for these types of models could come from barcodes, radio frequency identification (RFID) chips, global positioning systems (GPSs), enterprise resource planning (ERP) systems, electronic point of sales (EPOS) data and other data visualization tools with real-time information on quantities, qualities, stock levels, location and so on (Souza, 2014).

Predictive analytics

Predictive analytics asks questions about what is likely to happen, or what will happen, or what could happen. Predictive models and probability models could be included here as well as forecasting and statistical analysis (Appelbaum et al, 2018). Discovery analytics looks for key elements in the data, themes, patterns and correlations. Statistical tools such as linear regression, multiple regression and others could be categorized here. Lepenioti et al (2020) have conducted a systematic review of the literature with the main focus on prescriptive analytical models; however, they identify a number of predictive modes with the indication that in many cases the result of a predictive model is required as an input to a prescriptive model. Their investigation has not focused directly on the supply chain analytics field, but the identified models could also have a representation in this area. They are: probabilistic models, machine learning/data mining models and statistical analysis including linear, multiple, rank, logistics and multinomial regression, autoregressive integrated moving average (ARIMA), density estimation and support vector regression. Their study does not cover time series forecasting models, which are also a set of predictive models.

Prescriptive analytics

Prescriptive analytics is not only what happened and what is going to happen, but also what could be done (Wang et al, 2018). Prescriptive takes a step further from descriptive and predictive analytics by aiming to provide a solution and showing the outcome of these (Appelbaum et al, 2018). Souza (2014) and Nguyen et al (2018) classify analytical techniques as descriptive, predictive and prescriptive in a supply chain environment. Nguyen et al (2018) refer to techniques such as association, clustering, classification, semantic analysis, regression, forecasting, optimization, simulation and visualization. Wang et al (2016) note that prescriptive analytics could include multi-criteria decision-making, optimization and simulation. Predictive and prescriptive analytics may use similar techniques, but the purpose of this may be different. The more data used and the more varied the type of data considered for analysis, the more likely the technique is prescriptive (Appelbaum et al, 2018).

Lepenioti et al (2020) include in their review prescriptive analytics techniques such as probabilistic models, machine learning/data mining, mathematical programming, evolutionary computation, simulation and logic-based models.

The 'significant value creation' as depicted in Figure 1.1 is evaluated in Holsapple et al (2014) from a number of studies, and they summarize the rationale for implementing analytics, such as obtaining value from data, creating competitive advantage, improving an organization's performance, having a positive effect on a decision and generating knowledge.

Applications of business analytics

Last-mile optimization, predictive networking and capacity planning, customer value management, supply chain risk management, business-to-business (B2B) demand and supply chain forecasting, and real-time local intelligence are just a few applications of information technology (IT) and analytics adopted by DHL in logistics (Ittmann, 2015). We see a large number of applications of business analytics in the supply chain, still business analytics has been applied in a large number of areas. Just a few examples have been captured in Table 1.3.

Table 1.3 Some examples of business analytics applications

Area	Example
Accounting/ finance/budgeting	Examples include identification of customer service from feedback forms in different languages; prediction of risk in fraud and lending; credit history evaluation; customer behaviour prediction, and others. BA is also used in budgeting functions such as planning and evaluations (Bergmann et al, 2020). Nielsen (2018) indicates that some of the most important business analytics skills for management accountants are: holistic view, fact-based decisions, predictions and forecasting, visualization, creativity, communication and analytics skills. Other analytical models used in banking are: forecasting bank deposit rates, credibility of loan application, credit card fraud detection and others (Gupta, 2018).
Engineering	Product design and evaluation, automation, innovation; evaluation of feedback forms from product use and identification of repair activities; identification of components for remanufacture and many other applications.
Healthcare	The healthcare industry is producing a large volume of data on a daily bases, data that is generated at general practitioner centres, hospitals, pharmacies, health research centres and so on (Gupta, 2018). Data will be generated in a range of different formats and tailored analytics are expected for each of these types of data. Examples include: prediction (for example predict patients at high risk of developing diabetes); treatment identification with real-time information available; health risk identification; identification of critical case service; analysis of medical device data; drug information; advanced use of visualization and reporting of clinical performance metrics and others. Analytical models used in this industry may comprise the logistics regression model such as predicting patients at risk of becoming high-cost healthcare users, or decision tree models such as predicting breast cancer survivability, multiple regression models such as predicting length of stay in hospital (Gupta, 2018) or simulation models such as evaluating waiting times for receiving treatment.
Human resources	Analytics in human resources (HR) may include analysis on employee engagement, performance evaluation, training efficiency and effectiveness, employee satisfaction and retention, increase productivity and others (Bag, 2016).

(continued)

Table 1.3 (Continued)

Area	Example
Manufacturing	Production planning, capacity optimization, real-time scheduling, job-shop scheduling, order forecasting, predictive control, automated implementation of product design, automated operations, use of intelligent agents for decision-making and many others. BA in manufacturing can enhance innovation and generate new opportunities for value creation (Omar et al, 2019).
Marketing	Analytics are used by marketing organizations to determine the success of marketing campaigns, identify customer segmentation and other demographic studies to improve the decision-making process. Detailed information can be collected by web analytics to allow marketers to identify preferences, and therefore improve marketing campaigns (Plenert, 2014).
Service operations	Demand forecasting, demand planning, resource allocation, project management, operations planning, service optimization, risk analytics, visual analytics, real-time reporting; identifying customer service issues from feedback forms, emails, messages, social media and others.
Sports management	Predicting and reporting; player statistics, performance prediction.
Legal studies	Real-time visibility on legal documents; identification of the right legal talent; analysis of large sets of legal details; statistical analysis used to predict outcomes; graphical analysis used to identify patterns in numerical data.
Libraries	Storing and searching through a large volume of documents and information; identification of new resources based on previous search; text identification and many others.
Military	Rapid deployment of assets; strategic positioning and localized forecasting; identification of alternative routes; predictive maintenance for vehicle and other military resources; prediction of training and medical attention for soldiers.

Transportation	Transportation planning, capacity allocation, vehicle routing and scheduling; real-time routing and optimization; congestion identification; vehicle capacity planning; driver behaviour and others.
Supply chain management/ logistics	Analytics are used in production planning, forecasting, inventory control, optimization, capacity planning, resource allocation, distribution planning and others.
Port operations	Business analytics are used in improving port operations efficiency; for example, it can be used to shorten the time it takes to share information and documents between stakeholders (Zerbino et al, 2019); and to get information on incoming vessels to assist in port control (Tsou, 2019).
Bricks-and-mortar and online retailing	Customers address are used for developing optimum routing of deliveries; sales and demand data is used for forecasting, inventory control; EPOS data used for real-time planning. Analytics are used for: uncharacteristic pattern identification for internal and supply chain fraud; identification of opportunities for promotional activities; increased visibility between retailers and suppliers.
Education	Examples include skills identification for teacher hiring; statistical analysis on class performance; prediction of results.
Travel and tourism	Big data and data analytics are extensively used in the travel and tourism industry. Analytics in this area may consider travel preferences, travel history, travel locations, preferences on trip selection, aspects of customer experience and many others. A range of analytics are used from descriptive analytics on total number of passengers, for example, to predictive analytics, for example for personalized offers, predicting travel delays and many others (Gupta, 2018).

Spreadsheets are still the most popular when dealing with data, for various forms of data analysis, basic statistics, financial analysis, sales forecasting, reporting and others. Microsoft Excel is an application that is well used and understood, is easy to manipulate, and a user can easily transfer files to another user, and therefore a number of business analytics applications are carried out using Excel. Idoine et al (2019) have conducted a very detailed study of different data science and machine learning platforms supplied by a range of different vendors. Some of the software packages listed in their study are Alteryx, SAS, MathWorks, RapidMiner, IBM, Google, Microsoft, DataRobot, SAP, Domino, Databricks and others.

Other major software companies that offer prescriptive analytical tools in the area of optimizations are AIIMS, AMPL, Frontline, GMAS Lindo systems and many others (Sharda et al, 2017). Prescriptive analytical software such as simulation software are offered by Rockwell with the software ARENA discrete event simulation; Simio, again a discrete event simulation software for production planning and scheduling innovations; Simul8 used for planning, designing, optimization and reengineering of production, manufacturing and logistics systems; and Lanner Group comes with Witness simulation software. Vensim is another simulation software that works with system dynamics. Other companies that consider complex event processing products are Apache, Hitachi, Informatica and Tibco (Sharda et al, 2017). Stanton and Stanton (2020) have conducted a study evaluating the skills required by entry-level analysts and found that the software packages mentioned were as expected in the field of analytics and data science. Some of these include packages listed above as well as Apache (Flink, Hive, PIG, Spark and Storm), AWS, Azure, C/C++, Caffe, Cognos, Dataiku, Hadoop, Java, Matlab, Microsoft Excel and Access, Minitab, Python, R, Scala, Stata, SPSS, Tableau, Tensoflow, Teradata, Tibco, to name a few from their study. It would not be possible for one individual to master all of these; however, when one software package has been learnt it not as difficult to engage in learning another. Still, it is also not viable for one organization to implement all of these and make use of all of the software packages available on the market. The question therefore remains, which are the most appropriate software packages to implement and which skills to demand from a potential employee in the field of analytics and modelling? To further attempt to answer this question, it is more relevant to understand the business problem and evaluate the required skills and expectations from this angle. However, the goal may change with changes created by current environment and market expectations, therefore continuous training and awareness of new developments in the field should be a minimum expectation that organizations should have.

Key challenges for business analytics

The key challenge is to really understand what defines success in the area of business analytics. When can an organization claim to have had a successful BA implementation? What level of analytics is being carried out on a regular basis?

Implementing business analytics within an organization reflects on the implementation of the software technology, on the data currently available and the data required and on the measurement system used to evaluate performance. However, there is much more to the notion of analytics when this is being implemented. Abai et al (2019) comment on the fact that organizations use business intelligence technology but the use of analytics is not sufficiently implemented. Within their study, Abai et al identify 11 implementation activities: the identification of the problem and the opportunities offered; understanding the data to be used in the analysis; collecting the data; transforming this data ready for analysis; analysing the data; developing a model; evaluating the model; using the model; translating the outputs of the model; measuring the impact; and, the last activity, providing maintenance. Hawley (2016) says that cultural and environmental issues must also be taken into consideration. Kwon et al (2016, p 426) highlight the aspects of 'skills, technologies, applications and practices for continuous iterative explorations and investigation of past business performance' with the aim of understanding limitations and creating understanding for future improvements.

There is a growing understanding of business analytics tools and their usability, alongside the large volumes of data that are available to organizations. Many companies are investing in business analytics to help them extract more value from data internal to their organization as well as external. The process of extracting value from the available data and tools presents the real challenge to an organization.

Time, cost and energy are associated with understanding and adopting BA technologies within an organization. However, adopting BA technologies is not sufficient, they require appropriate personnel skills to manipulate these tools and technologies, they will need to link to current business technologies already implemented within an organization and be regularly maintained and updated, assessed and reassessed after implementation. Companies may focus their efforts on a particular department or section within their organization when considering implementing BA technologies. This approach may bring immediate benefits to this particular department within an organization, but its benefits may not be appreciated in full by the entire organization.

Successful outcomes of BA tools are directly linked to the accuracy of the data and the IT infrastructure available to work and manipulate the data available. Limited success within one section of an organization may have a negative impact on the adoption of BA technologies at a larger scale.

Oliveira et al (2012) refer to the fact that the implementation and use of BA may not show immediate results, and the benefits of using this type of technology should be measured with a time lag. Evaluating the limitations of the use of BA, Oliveira et al (2012) indicate that the impact of BA on performance does not only depend on supply chain performance measures, but also on the type of supply chain, the industry in question and its environment, and also the supply chain strategy.

Summary

Business analytics has been used by businesses for a number of years, and will continue to be used to bring new knowledge about the way in which operations are carried out, to improve the organization performance and improve efficiency, effectiveness and generate overall value. Analytics can be used in a descriptive way to provide understanding of what happened, what is happening, why it has happened based on the data available to an organization. Analytics can also be used in a predictive way to answer questions such as what is likely to happen. A descriptive as well as a predictive approach can be employed at the same time. More recently, organizations seek to understand what could happen if particular changes take place, therefore analytics could be used in a prescriptive way to provide understanding of what can be done.

Advances in technology, computer power and software development have facilitated the development and implementation of prescriptive analytics more and more. However, the business analytics arena is still dominated by descriptive and, to some extent, predictive analytics. Descriptive analytics offers valuable insights; however, they mainly refer to existing and/or past data with limited impact on complex predictions and complex decision making. There are continuous efforts from researchers and practitioners to develop, implement and test approaches that incorporate predictive and prescriptive elements, but more effort needs to be dedicated to this area of development.

To move towards a better understanding of what business analytics is, this chapter put forward a set of questions for practitioners and researchers that could be asked in relation with the topic. A large number of developed tools, and software packages available on the market have been highlighted to help with these issues, each with their benefits and limitations. Through this discussion it became apparent that dedicated skills and technologies need to be present for these tools to be implemented, used to their expected capacity and evaluated.

Business analytics can be applied in many areas within an organization as well as in many industries. Still, the skills to approach, develop and implement analytics in a practical setting are similar regardless of the industry or the individual sections within an organization in which it is applied. Learning to work with business analytics tools is not only one of the most powerful skill to acquire, but one that is most sought after by the industry.

References

Abai, N H Z, Yahaya, J, Deraman, A, Hamdan, A R, Mansor, Z and Jusoh, Y Y (2019) Integrating business intelligence and analytics in managing public sector performance: An empirical study, *International Journal on Advanced Science, Engineering and Information Technology*, 9(1), 172–80

Al-Haddad, S, Thorne, B, Ahmed, V and Sause, W (2019) Teaching information technology alongside business analytics: Case study, *Journal of Education for Business*, 94(2), 92–100

Appelbaum, D A, Kogan, A and Vasarhelyi, M A (2018) Analytical procedures in external auditing: A comprehensive literature survey and framework for external audit analytics, *Journal of Accounting Literature*, 40, 83–101, doi:10.1016/j.acclit.2018.01.001

Bag, D (2016) *Business Analytics*, 1st edn, Taylor and Francis, Abingdon

Barbosa, M W, Ladeira, M B and de la Calle Vicente, A (2017) An analysis of international coauthorship networks in the supply chain analytics research area, *Scientometrics*, 111(3), 1703–31

Bergmann, M, Brück, C, Knauer, T and Schwering, A (2020) Digitization of the budgeting process: Determinants of the use of business analytics and its effect on satisfaction with the budgeting process, *Journal of Management Control*, doi:10.1007/s00187-019-00291-y

Gupta, D (2018) *Applied Analytics Through Case Studies Using SAS and R: Implementing predictive models and machine learning techniques*, Apress, Berkeley, CA, doi:10.1007/978-1-4842-3525-6

Hawley, D (2016) Implementing business analytics within the supply chain: Success and fault factors, *Electronic Journal of Information Systems Evaluation*, 19(2), 112

Holsapple, C, Lee-Post, A and Pakath, R (2014) A unified foundation for business analytics, *Decision Support Systems*, 64, 130–141, doi:10.1016/j.dss.2014.05.013

Idoine, C, Krensky, P, Brethenoux, E and Linden, A (2019) Magic quadrant for data science and machine learning platforms, Gartner, January

Ittmann, H W (2015) The impact of big data and business analytics on supply chain management, *Journal of Transport and Supply Chain Management*, 9(1), e1–e9

Kwon, I-W G, Kim, S-H and Martin, D G (2016) Healthcare supply chain management: Strategic areas for quality and financial improvement, *Technological Forecasting and Social Change*, 113, Part B, 422–428

Lepenioti, K, Bousdekis, A, Apostolou, D and Mentzas, G (2020) Prescriptive analytics: Literature review and research challenges, *International Journal of Information Management*, 50, 57–70, doi:10.1016/j.ijinfomgt.2019.04.003

Neely, A and Jarrar, Y (2004) Extracting value from data: The performance planning value chain, *Business Process Management Journal*, 10(5), 506–09

Nguyen, T, Zhou, L, Spiegler, V, Ieromonachou, P and Lin, Y (2018) Big data analytics in supply chain management: A state-of-the-art literature review, *Computers and Operations Research*, 98, 254–64

Nielsen, S (2018) Reflections on the applicability of business analytics for management accounting – and future perspectives for the accountant, *Journal of Accounting and Organizational Change*, 14(2), 167–187, doi: 10.1108/JAOC-11-2014-0056

Oliveira, M P V d, McCormack, K and Trkman, P (2012) Business analytics in supply chains: The contingent effect of business process maturity, *Expert Systems with Applications*, 39(5), 5488–98

Omar, Y M, Minoufekr, M and Plapper, P (2019) Business analytics in manufacturing: Current trends, challenges and pathway to market leadership, *Operations Research Perspectives*, 6, doi:10.1016/j.orp.2019.100127

Plenert, G J (2014) *Supply Chain Optimization Through Segmentation and Analytics*, 1st edn, vol 48, CRC Press, Boca Raton, Florida

Sharda, R, Delen, D, Turban, E, Aronson, J E, Liang, T-P and King, D (2017) *Business Intelligence, Analytics, and Data Science: A managerial perspective*, 4th edn, Pearson Harlow

Sivarajah, U, Kamal, M M, Irani, Z and Weerakkody, V (2017) Critical analysis of big data challenges and analytical methods, *Journal of Business Research*, 70, 263–86

Souza, G C (2014) Supply chain analytics, *Business Horizons*, 57, 595–605

Stanton, W W and Stanton, A D (2020) Helping business students acquire the skills needed for a career in analytics: A comprehensive industry assessment of entry-level requirements, *Decision Sciences Journal of Innovative Education*, 18(1), 138–65, doi:10.1111/dsji.12199

Sukhobokov, A A (2018) Business analytics and AGI in corporate management systems, *Procedia Computer Science*, 145, 533–44

Tsou, M C (2019) Big data analytics of safety assessment for a port of entry: A case study in Keelung Harbor, *Proceedings of the Institution of Mechanical Engineers Part M: Journal of Engineering for the Maritime Environment*, 233 (4), 1260–75, doi:10.1177/1475090218805245

Wang, G, Gunasekaran, A, Ngai, E W T and Papadopoulos, T (2016) Big data analytics in logistics and supply chain management: Certain investigations for research and applications, *International Journal of Production Economics*, 176, 98–110

Wang, C-H, Cheng, H-Y and Deng, Y-T (2018) Using Bayesian belief network and time-series model to conduct prescriptive and predictive analytics for computer industries, *Computers and Industrial Engineering*, 115, 486–94

Wang, S, Yeoh, W, Richards, G, Wong, S F and Chang, Y (2019) Harnessing business analytics value through organizational absorptive capacity, *Information and Management*, doi:10.1016/j.im.2019.02.007

Zerbino, P, Aloini, D, Dulmin, R and Mininno, V (2019) Towards analytics-enabled efficiency improvements in maritime transportation: A case study in a Mediterranean port, *Sustainability (Switzerland)*, 11(16), doi:10.3390/su11164473

Bibliography

Evans, J R (2017) *Business Analytics: Methods, models, and decisions*, 2nd edn, Pearson, Boston

Laursen, G H N and Thorlund, J (2010, 2017) *Business Analytics: Taking business intelligence beyond reporting*, Wiley, Hoboken, NJ

Tripathi, S S (2016) *Learn Business Analytics in Six Steps Using SAS and R: A practical, step-by-step guide to learning business analytics*, 1st edn, Apress, Berkeley, CA

The role of modelling in the supply chain 02

LEARNING OBJECTIVES

- Provide an understanding of supply chains as systems.
- Ask and reflect on questions related to supply chain modelling.
- Understand supply chain complexity and issues associated with modelling complexity.
- Identify steps in developing a supply chain model.

Introduction

In an attempt to look at the role of modelling in the supply chain, first, it is relevant to understand what supply chains are and how are they defined. Supply chains are characterized by complex structures and therefore understanding what defines complexity within a system is considered next. To approach supply chains from a modelling and analysis point of view, the chapter considers what are models, what is their role and what types of models have been successfully applied in practice. Modelling complex systems is a challenging task, but a number of benefits have been associated with this approach. However, this is not to say that modelling supply chain systems is free of challenge, limitations and a large number of assumptions. Understanding the way in which models are created in the context of supply chains would allow users to be able to analyse the data, processes and operations as a result of the modelling task.

As modelling appears a challenging and, in many cases, an abstract notion, it helps to understand what types of questions could be asked in relation to this area of investigation. Some of the questions in Table 2.1 offer opportunities for reflection and are believed to be of benefit to practitioners, analysts, business managers, IT managers, software engineers, modellers, supply chain managers and researchers.

Table 2.1 Sample of supply chain modelling questions

Questions	Relevant to:
What is a supply chain system? What is a supply chain network? What defines complexity within a supply chain system? How can 'good' and 'bad' complexity be identified in a supply chain? What are the key elements that need to be considered when modelling a supply chain system?	Researchers, practitioners, organizations
What data is required to model a particular supply chain system? What data is available to model a supply chain system? Who is going to benefit from modelling a particular supply chain system?	Organizations: strategic managers, operational managers, process managers, data warehouse managers, IT managers
How can a supply chain model be to verified and validated?	Organizations: business analysis, supply chain modelling managers Researchers: in the areas of operations management, operational research and supply chain modelling and analytics
What software packages and ERP system are required to model and implement the developed supply chain system?	Organizations: business analysis, supply chain modelling managers, IT managers, finance managers, software vendors Researchers: in the areas of operations management, operational research, supply chain modelling and business analytics

(continued)

Table 2.1 (Continued)

Questions	Relevant to:
What skills are required to model, verify, validate and implement a supply chain model?	Organizations: supply chain modelling managers, business analysts, HR managers Researchers: in the areas of operations management, operational research, business analytics and supply chain
What are the critical success factors for implementing a supply chain model? What leads to post-implementation success?	Organizations: strategic, tactical and operations managers Researchers: in the areas of operational research, operational management, strategic management, behaviour science, logistics and supply chain
What defines an overall objective when modelling a supply chain system?	Researchers: in the areas of operational research, operational management, logistics and supply chain and performance management
How can supply chain modelling be used to promote innovation?	Organizations: strategic, tactical and operations managers, engineers Researchers: in the areas of operational research, operational management, strategic management, behaviour science, logistics and supply chain

Defining the supply chain

The supply chain system and supply chain management have received considerable attention and various definitions have been given to capture the complexity of the activities that form part of the supply chain. When defining the supply chain, it is relevant to take into consideration the characteristics of the physical flow, data and information flow, service flow, financial flow, knowledge and personnel or social flow. When referring to the supply chain as a system, this should not be reduced to the physical flow, nor should this be considered in isolation. The physical flow captures the movement of goods from raw materials, to semi-finished and finished products, to their distribution to the final consumer using the most efficient form of raw material selection, transformation and transportation using the most effective form of delivery of the final product. However, the physical flow of goods does not stop here. Products continue to influence the original supply chain, a different supply chain or the environment after reaching the final consumers, and after being used by them for a period of time. Products may be returned under various functions; they may be recalled, repaired, reused, reassembled, remanufactured, repurposed or destroyed. Together with the physical flow of a product in the supply chain, data and information flow plays a key role in facilitating the physical flow of products along its complex structure. Currently, a number of challenges surround the issue of data, some of which are further debated in the following chapters, with specific reference in Chapter 4. The data is not only internal to an organization forming the structure of the supply chain, but also external to an organization. Data is not only internal to the supply chain system, but also external to the system. Through means of analytics, this data has the potential to be transformed into information and knowledge that not only supports the physical flow of goods in the supply chain, but also supports the other key flows, such as services, financial flow and social flow in the systems.

A number of representations of supply chains have been attempted over the years, with the most common representation being the linear view where the chains is formed by linking suppliers with intermediaries, and these with manufacturers, and these with wholesalers, and these with retailers, and so on. This linear representation has its benefits for understanding the basic flow and characteristics of a supply chain, but it is oversimplified. In an attempt to indicate complexity and provide a more detailed representation, a general view of a supply chain system could be represented as in Figure 2.1. Semi-finished and finished products flow in the supply chain network but they do not necessarily visit all organizations in this network.

The flow of information in the represented network has the same characteristic. Some organizations are clearly connected through IT systems that communicate with each other to facilitate the flow of information, but others do not offer the opportunity for sharing the required information, or do not have the IT capability to do this. There are also data security issues that some organizations in the supply chain are not able to resolve, therefore access to data and more access to reliable data is not always possible.

Defining the supply chain has challenged the academic and professional community for a number of years. Are we talking about a system, or are we

Figure 2.1 A general representation of supply chain flows

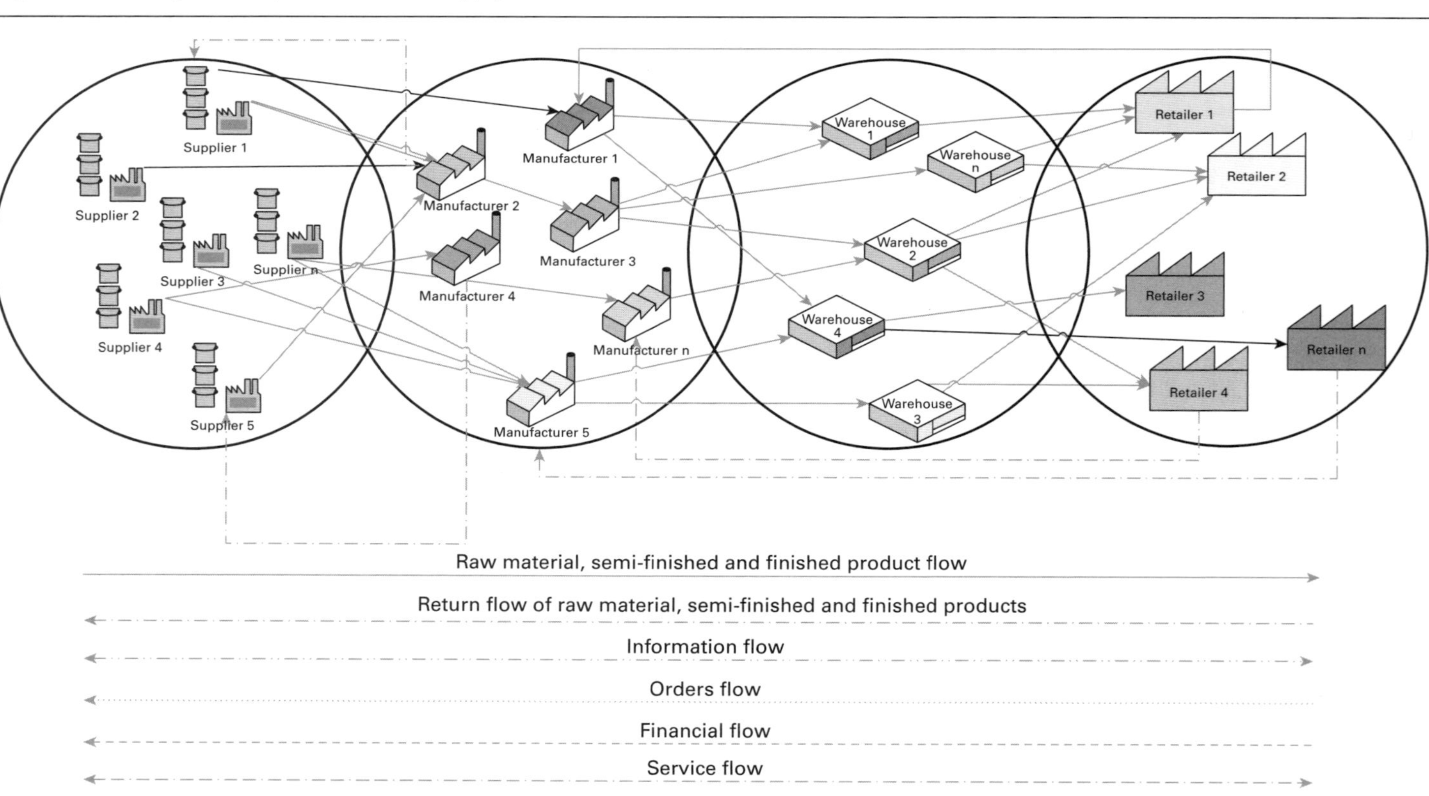

engaging in looking at the supply chain as a network? 'Systems' as well as 'networks' as terms have been used in associations with the supply chain, and especially in the context of modelling and analysis. Both of these are acceptable, but there are some differences when aiming to approach the supply chain from a system point of view or from a network point of view. In published literature, in the context of supply chain, these terms may have also been used interchangeably where establishing the difference between the two has not necessarily been considered essential. Establishing a clear distinction of a system within reality helps to provide an understanding, description and to bring some grouping or separation from the multitude of elements ready for selection, to establish specific measures and to facilitate clearer analysis and evaluation. Dekkers (2017, p 16) defines applied systems theory and characterizes the system as 'consisting of elements discernible within the total reality (universe), defined by the aim of the investigator'. From a supply chain point of view, we would like to see the 'discernible elements' as individual entities/organizations where the aim is to form the supply chain. Dekkers goes further with the definition of a system and adds 'all these elements have at least one relationship with other elements within the system and may have relationships with other elements within total reality'.

To be able to model and analyse the supply chain from a system point of view, the following assumptions should be considered: entities (or organizations in this case) for which data is available and known, the links between these entities (organizations) are well established and their position and characteristics in the chain are known. In this case the analysis could be conducted on the whole system. This system may be influenced by external factors, other entities, or the environment, but the analysis could still be carried out in its entirety on the selected system. A system as it is captured for analysis from a supply chain point of view can in fact be a sub-system of a bigger system and depends on where the boundary is being considered.

However, supply chains could also be analysed from a network point of view, where there are system-related elements, clearly defined and linked, and there are entities that are known, or unknown, that still have an influence on the known system (see Figure 2.2).

Using this definition, a group of supply chain systems could form a network or be parts of a bigger network. Networks are seen as scale-free (Dekkers, 2017), with potentially a large number of nodes, some of which form strong connections.

Lambert and Cooper (2000) refer to the structural dimension of the supply chain network for describing, analysing and managing the supply chain

Figure 2.2 Supply chain system with internal and external entities and links

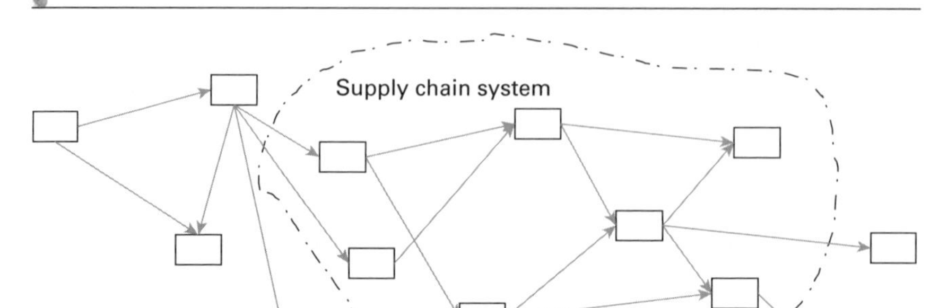

following a horizontal structure, a vertical structure and a horizontal position in the chain of the focal company under investigation. In their definition, the horizontal structure captures the number of tiers across the chain, where a long chain is represented by many tiers. Within the vertical structure, they refer to the number of customers or suppliers within each tier. The last structural dimension in this case is the horizontal position of the company under investigation along the supply chain; this may be closer to the supplier end, or may be closer to the retailer/customer end of the chain.

Mentzer et al (2001) identify three distinct classifications of supply chains. The first is in the form of the direct supply chain. The analysis regarding a direct supply chain would consider the company under investigation, its supplier and customer involved in carrying out operations and representing the flow of products, information, services and the financial flow. The second type of supply chain identified would be the extended chain, where the supplier's supplier and customer's customer form part of the analysis. The last category identified in this case is the ultimate supply chain, where all organizations forming part of the system as well as the flows associated, such as products, services, financial flow and information flow, are captured into the analysis. However, this understanding has been put forward for a number of years, and examples capturing an 'ultimate supply chain' are limited in their approach from a modelling and analysis point of view. The term 'long supply chain' is used (Jain and Benyoucef, 2008), as is 'end-to-end supply chain' (Chilmon and Tipi, 2014). These studies stress the importance of providing an analysis of the entire supply chain if this is possible, or a pre-established end-to-end supply chain system.

The structure of a supply chain is continuously changing, and this is to recognize some of the external factors. For example, during the COVID-19 pandemic we saw changes in the structure of supply chains, where many chains operating globally had to change to more local approaches, from a supplier perspective, or a manufacturing or distribution perspective. The idea of creating supply chains that have the capability to adapt to current environmental changes is not new, and a number of studies have conceptualized the idea of complex adaptive supply chains (Choi et al, 2001; Choi and Krause, 2006; Hearnshaw and Wilson, 2013).

The supply chain as a complex system

From the number of parts that form the supply chain, the number of activities that take place in managing and coordinating the chain, the value of information that affects the supply chain as well as the volume of information, data, knowledge, products and services that are generated as a result of supply chain operations, it is evident that the supply chain has all the characteristics of a complex system. Supply chains are increasingly complex, running from domestic to international operations, from horizontal to vertical integration.

There is a need to generate new products, develop new processes, find new ways to deliver products, redevelop designs for products and processes that would attract customers and many other initiatives that will all trigger activities in the chain that would lead to added complexity. The interrogative denunciations in this instance would be to what extent is complexity in the supply chain to be encouraged, allowed or repressed? A decision of this nature on one section of the supply chain does not only have an immediate effect on this section of the chain, instead it will have an effect to some degree on the entire chain. Some decisions in a particular part of the chain are propagated further in the chain, having an amplified effect (for example, the amplification of orders or bullwhip effect in the supply chain) on certain parts of the chain, or a continuous resonant effect (see, for example, the resonant effect of inventory) in sections of the chain, and probably no effect or very small effect on other parts. Being able to understand the effect a decision on one section of a chain has on the entire supply chain network, could bring significant benefits to organizations forming part of the supply chain. This understanding could be achieved through modelling that would comprise understanding the system under investigation, understanding the extended network, understanding the environment, understanding the modelling and being able to analyse and implement the best scenario.

To what extent would added complexity in the chain lead to value added? 'Value' in this respect would need to be looked at from an economic, environmental as well as social aspect. What actions would need to be put in place to understand what complexity should be added that would lead to added value in the chain? Following this enquiry, we would then ask, will this be possible? Who will have the responsibility in the chain to decide on the level of complexity that is 'allowed'?

This question will lead us to the understanding that this may not possible, as there is no one party that will have overall responsibility in the chain to allow for this level of control. Therefore, the challenge facing supply chain systems is even more evident as individual organizations forming part of the system would aim to gain efficiencies, and financial and operational advantages with a localized context, without having the knowledge or understanding of how this aspect would impact the overall system under investigation.

Over the years, a number of researchers have explored the need to tackle complexity in the supply chain and provide more information about what needs to be put in place to manage this complexity.

Going back to the work of Weaver (1948), we see that system complexity was captured in three ways: systems of organized simplicity; systems of disorganized complexity; and systems of organized complexity. The supply chain complexity triangle is identified in Wilding (1998) who also specifies three elements: deterministic chaos, parallel interactions and demand amplifications. As highlighted in this case, avoiding complexity may be very difficult where complexity could be seen as a threat. Blackhust et al (2004) also look at supply chain complexity and bring forward elements such as a large number of interlinked participants in the supply chain; each participant can be a member of other supply chains; the dynamic and uncertain nature of the supply chain; and each participant can have different objectives.

Looking at complexity from a supply base perspective, Choi and Krause (2006) put forward three dimensions: the number of suppliers in the base, the level of differentiation of these suppliers, and the level of connectivity, or interrelationship, between these. What is relevant to capture regarding aspects of complexity is: the total number of entities available, the characteristics associated with these entities, and the links between them.

Looking at supply chain complexity, Milgate (2001) considers uncertainty, technological intricacy and organizational systems as forming the supply chain complexity definition. There is also the debate of complexity and uncertainty in the chain. According to Milgate (2001), complexity is related more to the type and the level of interaction in the system, whereas uncertainty is linked to the variations present in the system. These variations could take the form of 'late deliveries', 'product quality', 'unexpected variations in demand', and so on.

It is also relevant to recognize that there is a clear distinction between complicated and complex structures. A complicated structure could be formed from a very large number of entities, but the connection between all is fixed and clearly established. However, looking at the complex system, according to Pathak et al (2007), this is represented by non-linear dynamic interactions of individual parts. They go further and argue that a complicated system could be analysed as an addition of its individual components; however, this is no longer the case when analysing a complex system. The behaviour of a complex system cannot be predicted from analysing the behaviour of its individual parts. Therefore, in modelling and analysing a supply chain system, it is imperative to capture the entire system under analysis and understand the driver of complexity that characterized the selected system under investigation. Bozarth et al (2009) capture elements of the system as detail complexity and dynamic complexity. In their definition the detail complexity is represented by the individual number of elements or components that form the system, where the dynamic complexity comes from the unpredictability of the system output based on a given set of inputs, unpredictability that will be driven by the many links that form the system.

Flood and Jackson (1991) (referenced in Skyttner 1996, p 66) attribute particular characteristics to a complex system that refer to the number of elements incorporated in the system and the interaction between them; with undetermined attributes where the interaction between them is loosely organized. These elements could have probabilistic behaviour and they could evolve over time with subsystems that are purposeful and generate their own goal. The system can also be subject to changes in the environment.

Following from the characteristics presented above, complex supply chain systems could have the following features:

- **A supply chain is formed of a large number of entities.** Following the physical flow (see Figure 2.1) we see a large number of organizations contributing to the formation of the supply chain network from a diverse range of suppliers, to service providers, to manufacturers or particular operations, to distributors, to retailers, to final customers, to return organizations and so on. A large number of entities are also observed internally by an organization from the number of different departments, products type, process types, and services to the type of decisions that are taking place on a regular basis. A large number of entities are also present from the information flow angle that comprises the large number of transitions taking place every day, to the number of IT systems used to monitor, evaluate, control, process and analyse information.

- **A supply chain system/network has many interactions between elements.** There are continuous interactions between different organizations seen as external interactions in the supply chain and between different internal functions to an organization. These interactions could have static as well as dynamic behaviour. Links are formed between new members in the chain and these are continuously changing, such as new contracts are being set, new channels of distribution being established, or new materials being purchased, or relations could be dissolved, where no new materials are being delivered and so on.
- **The attributes in supply chain interactions are predetermined.** It can be argued that the attributes of the elements in the supply chain interactions are in fact predetermined, where individual functions in the chain are well characterized (for example, a supplier of particular raw materials will retain this function, but they may also take on the function of a manufacturer, or the function of storage, or provide a transport service).
- **There are many interactions between elements.** In a supply chain environment, it is also relevant to say that there are many interactions between these elements, and they follow an organized approach, but this could change based on new opportunities, or new changes in the market, or new challenges as a result of new environmental conditions. It may not always be evident that the interaction between elements is loosely organized, or whether the behaviour is always characterized by probabilistic behaviours, but this could be the case with many examples in the supply chain.
- **The supply chain system evolves over a period of time.** New products are introduced, new suppliers are approached, new customers are continuously acquired, new markets are explored, new IT systems are put in place to support operations and so on.
- **A supply chain is formed of a number of sub-systems.** These include suppliers, manufacturers with their internal supply chain, distributors, retailers with their internal supply chain, customers, each with their own set of goals. All these entities or echelons in the supply chain may be seen and analysed as sub-systems in the chain. For example, a manufacturer could have its own sales and distribution unit, as well as be formed by a

number of manufacturing plants. Still, the manufacturer could be seen as the sub-system in the supply chain system, where even this sub-system could be formed by a number of individual sub-systems.

- **The supply chain is subject to behavioural influences and is largely open to the environment.** For example, if a product is no longer successful in the market, this will be replaced, and where new product characteristics are put in place new processes and new suppliers are required. This change will then generate a change to the structure and characteristics of the supply chain under investigation.

There are, however, other forms of complexity within a supply chain system on top of structural complexity. Manuj and Sahin (2011) also look at the terminology of supply chain complexity and they make the distinction between supply chain complexity and supply chain decision-making complexity. For the supply chain complexity, they bring elements such as structure, type, volume of activities, transactions that take place in operating a supply chain system. Supply chain decision-making complexity refers to the complexity of taking decisions in operating and managing the system. This aspect may include the way in which data is being collected, how the problem at hand is defined, and maybe how a solution can be implemented.

Other categorizations of supply chain complexity are presented by Serdarasan (2013), grouped as static, dynamic and decision-making complexity. Under each of these types, the author brings three categories such as: internal, supply/demand interface and external drivers. The static complexity driver refers to the structural complexity as well as process interactions and new technologies, and the dynamic drive makes more reference to supply chain management issues such as inappropriate forecasting, demand amplification and lack of process synchronization aspects that could benefit from a modelling solution. The decision-making complexity aspects refers to the IT systems used and different decision-making processes and information gaps, that again could be approached with business analytics tools and techniques.

Turner et al (2018) talk about the structural complexity in a supply chain as well as socio-political and emergent supply chain complexity. The complexity dimensions in their case take responses not from a reduction or accommodation of complexity point of view, but from dimensions such as planning and control, relationship development and flexibility for all three categories of structural, socio-political and emergent supply chain complexity. In their studies they have deduced that in many cases managers are dealing with the complexity present within managing the supply chain rather than reducing this complexity and employ project management solutions to explore the issue of complexity in the supply chain.

When dealing with complexity, organization may decide to accommodate it or aim to reduce it. They may want to investigate whether complexity will bring benefits or if it is damaging to the business. According to Turner et al (2018), it is relevant to enquire what type of response a company should consider, such as a plan and control type approach, or forming a relational response or looking for a flexible approach.

What decision-making tools and techniques could be employed to deal with the level of complexity? Are modelling and modelling tools and techniques a solution to move forward in dealing with supply chain complexity?

Supply chain models and modelling

Under the operational research/management science (OR/MS) area, a number of models relevant to supply chain have been developed over the years.

Mitroff et al (1974) considered different representations of a system from a problem-solving point of view. Their approach starts with the fact that there is a reality, from which a conceptual model can be developed. From the conceptual model, using modelling techniques a scientific model can be developed. Solving the scientific model, a set of solutions can be identified, using different scenarios. If these are appropriate, they can be implemented in the real situation. There are different variations to these steps, where the interest is to look at developing the conceptual model as well as developing the scientific model through modelling to provide solutions.

There are models that can be used in a particular situation. These could be time dependant models, for example the facility location model, which aims to identify the location of a new warehouse or a new manufacturing location. This type of model will only be used once in the modelling process. One other example is the allocation model that aims to identify whether building a new airport is a sustainable decision. Again, this type of model may only be required to be used once from a modelling perspective.

There are also models developed that are to be used on a continuous basis, such as a forecasting model, that are used at organizational level, for example for the number of products to be produced, or are used at different points in the supply chain, such as at the retail site, supplier site, and/or at a distribution centre.

Decisions are taken based on the results from the analysis the models produce, where these decisions could result in changes in the supply chain system structure. The transportation model, the network model, the gravity model all have the power to influence structural changes such as where the new location of a warehouse could be, or how the distribution network could be changed, and so on.

Models can be developed as a direct representation of circumstances with the scope to understand, evaluate, analyse and/or control this reality. They can also be purposely built to anticipate different events, or react to unpredictable disturbances as well as simply mirroring the flow of different activities and their level of interaction.

A model is defined by Pidd (2010, p 10) as an 'extended and explicit representation of part of reality as seen by people who wish to use that model to understand, to change, to manage and to control that part of reality'. The roles of models identified within this definition are to understand, to change, to manage and to control a part of reality where the design or redesign of this part of reality can be included in the change function. Pidd (2010) identifies four types of models depending on the degree to which they are used and the amount of human interaction involved in making the model ready for each use. They are: decision automation models, routine decision support models, system investigation and improvement models, and models to provide insight for debate. This further extends the role of modelling, as models are not only used to investigate and for scenario analysis – there are also dedicated models that can have automated functions or models used for routine decisions where human interaction is required less than for models used for investigation.

Morris (1967) in his work 'On the art of modeling' notes that skills in modelling require a sensitive and selective perception of management situations, where this depends on the conceptual structure one has available. He also summarizes and considers that relatedness, transparency, robustness, fertility and ease of enrichment are useful characteristics for models. Morris (1967) defines these characteristics as follows:

- Relatedness is where previously demonstrated theorems and results can be reflected in the model.
- Transparency refers to how obvious the interpretation is on the model and how immediate is its intuitive confirmation.
- Robustness is seen as the sensitivity of a model to changes in assumptions.
- Fertility is how rich is the variety of deductive consequences that the model produces.

- Ease of enrichment refers to what difficulties are encountered by attempts to further develop and elaborate the model in various directions.

This clarifies that models should have particular characteristics, where it is important to understand whether current supply chain models reach these characteristics or whether they have been developed with the scope to understand reality and develop theory.

The use of computer modelling in the supply chain

Models have different natures, such as visual models, paper-based models, mathematical models or simulation models, but this section refers only to computer-based models relevant to supply chains.

Models and modelling bring a number of benefits to organizations in providing efficiency in operations, improving performances and generating innovation.

A few studies have looked to evaluate and provide different classifications for supply chain models. Multi-stage models for supply chain design and analysis are reviewed and divided by Beamon (1998) into four categories: deterministic analytical models, stochastic analytical models, economic models, and simulation models. Most papers reviewed here consider models under the first two categories, with only a few coming under the economic and simulation headings. It is also relevant to note that when investigating the issues of supply chain models and their characteristics, Beamon (1998) assess the performance measures and their decision variables, and further details of this aspect are given in Chapter 4 of this book. Riddalls et al (2000) review various mathematical methods used to analyse supply chains and arrange them into the following categories: continuous time differential equations, discrete time differential equations models, discrete event simulations systems, and operational research techniques. They conclude that while operational research techniques are useful in providing solutions to local tactical problems, the impact of these solutions on the whole supply chain can only be assessed using dynamic simulations.

Shapiro (2001) splits analytical IT models used in supply chain modelling into descriptive and normative or optimization models. The descriptive models include the following: forecasting models, cost relationships, resource utilization relationships and simulation models. The optimization models are those using mathematical programming models.

Bertrand and Fransoo (2002), discussing methodologies used in quantitative modelling research, arrive at the following classification of existing models: axiomatic models and empirical models. They go further and define descriptive and normative aspects for each type of model. Each of their types of models receives valuable characteristics based on the model developed by Mitroff et al (1974), which starts by describing the entire modelling process with its specific four phases: conceptualization, modelling, model solving and implementation. Therefore, for an axiomatic descriptive model the modeller takes a conceptual model from literature and provides a scientific model through the modelling. An axiomatic normative model goes through the modelling and model-solving phases, which seek to provide a solution for the scientific model. An empirical deterministic model follows the cycle of conceptualization, modelling and validation. However, for the empirical normative model, the entire cycle is considered, such as conceptualization of a model from a real problem, where through modelling a scientific model is identified and a solution provided through model solving. This solution is then implemented in a real situation.

Min and Zhou (2002) develop a taxonomy of supply chain models where they categorize models as deterministic models, stochastic models, hybrid models and IT driven models. Models considering single and multiple objectives are grouped under the deterministic models, whereas optimal control theory and dynamic models come under the stochastic models' group. Hybrid models are considered models with both deterministic and stochastic elements and they include inventory-theoretic models and simulation models. The IT driven models include vendor management systems (VMS), enterprise resource planning (ERP) models and geographical information systems (GIS) models. Together with this classification of models, they also identify the following components as key to the modelling of supply chains: supply chain drivers, supply chain constrains and supply chain variables. It is relevant to list these elements to be able to compare them with previous studies and to be able to identify similarities with other studies where supply chain models have been developed and used. Under the supply chain drivers Min and Zhou (2002) identify the customer service initiatives, monetary values, information/knowledge transactions and risk elements. For supply chain constraints the following three have been identified: capacity, service compliance and the extent of demand. The list presented for the decision variables in the supply chain is significantly longer than the list identified in Beamon (1998), however this has not been linked to specific models in the literature. This list comprises variables such as location, allocation, network structuring, number of facilities and equipment, number of stages (echelons), service sequence, volume, inventory level, size of workforce and the extent of outsourcing.

Simulation, heuristics and optimization are also identified in Slats et al (1995) as modelling approaches applied in the logistics field. They mention that the logistics models in use are mainly simulation and optimization models, where the analytical models encompass too many assumptions and reduce or ignore many uncertainties. Modelling approaches considered in Blackhurst et al (2004) are classified in optimization models versus pure modelling models. The optimization approaches considered here include linear programming, integer programming and nonlinear programming where the pure modelling approaches consider simulation and network-based approaches. However, Blackhurst et al (2004) acknowledge that drawbacks exist for each modelling approach and their preferred method is the network-based approach.

Following a detailed literature investigation in supply chain management from 2004 to 2008, Kabak and Ulengin (2011) provide a classification of the type of modelling present in the reviewed papers and their supply chain environments. They identify six types of modelling: deterministic single objective, deterministic multiple objectives, stochastic, hybrid (deterministic and stochastic), information technology driven models and fuzzy set theory models. The most-used type of modelling has been identified here as the deterministic single objective. This review also concludes that the fuzzy set, IT driven and hybrid models have not been used frequently. Brandenburg et al (2014) provide a classification of models used in the analysis of sustainable supply chain management research. The mathematical programming methods observed earlier are now receiving more attention and comprise specific methods such as linear programming, mixed-integer linear programming, goal programming, dynamic programming, queuing models, nonlinear programming. Artificial intelligence techniques comprising of Petri net, case-based reasoning, Bayesian network, fuzzy logic, neural networks and rough set modelling are classed here as heuristic methods. Meta-heuristics modelling has captured models such as genetic algorithm, simulated annealing, differential evolution, particle swarm optimization, ant colony optimization and greedy randomized adaptive search procedure.

Table 2.2 shows a selection of categories of models used so far in the literature. There is no clear consistency between models and their classification identified so far; however, deterministic single objectives optimization models appear in most of these classifications.

Table 2.2 Categories of computer-based models in the supply chain

Author (year)	Classification of models
Beamon (1998)	Deterministic analytical models Stochastic analytical models Economic models Simulation models
Riddalls et al (2000)	Continuous time differential equations Discrete time differential equations models Discrete event simulations systems Operational research techniques
Shapiro (2001)	Descriptive models Normative or optimization models
Bertrand and Fransoo (2002)	Axiomatic models Empirical models
Min and Zhou (2002)	Deterministic models Stochastic models Hybrid models IT driven models
Blackhurst et al (2004)	Optimization models Network based models Simulation models
Kabak and Ulengin (2011)	Deterministic single objective Deterministic multiple objectives Stochastic models Hybrid (deterministic and stochastic) IT driven models Fuzzy set theory models
Brandenburg et al (2014)	Mathematical programming methods (single and multi-objectives) Simulation method (spreadsheet, system dynamics, discrete event and business game) Heuristic method (simple heuristic, artificial intelligence, meta-heuristics) Hybrid models Analytical models (multi-criteria decision making, game theory and systemic models)

A large number of models have been developed over the years in the area of supply chain. However, the problem to be solved in the supply chain determines which model is to be employed for analysis. One key element in this process is how accurately the problem has been identified for the models and analytics to be effective (Abai et al, 2019).

Developing a model and operating with a supply chain model will require a number of stages, as described in Table 2.3. This would start with understanding the reality and identifying the problem at hand. As soon as the need to model has been identified, a conceptualization stage is required. The conceptualization stage is seen as the preparation stage, where a number of tasks

Table 2.3 Stages in developing a supply chain model

Stages	Steps	Requirements
Understanding reality, identifying the problem	1	Evaluate the current situation
	2	Identify elements leading to effectiveness, efficiency, sustainability and resilience
	3	Identify problems/issues that hinder effectiveness, efficiency, sustainability and resilience
	4	Group problem areas in categories
Conceptualization stage	1	Identify the need to model, where a new model, or a model that already exists, is required A model may exist; however, its objectives may need readjusting due to current business needs
	2	Define the structure of the supply chain under investigation (single echelon, dyadic structure, multi echelon linear structure, convergent structure, divergent structure, network structure within a predefined system, interlinked supply chain structures, closed or open loop supply chains)
	3	Understand the characteristics of the supply chain system (level of complexity, the number of elements that form the system, the interaction between these elements, the dynamic characteristics of the system)
	4	Consider the type of investigation. This may involve descriptive, predictive or prescriptive approaches
	5	Define the supply chain's objectives. Based on the supply chain structure already identified, the objectives may be set for the entire chain, and/or may be set to individual stages in the chain (vertically within an organization followed by a horizontal approach within the supply chain)
	6	Identify the organizational performance measures and the overall supply chain performance measures

(continued)

Table 2.3 (Continued)

Stages	Steps	Requirements
Modelling stage	1	Determine the model's scope
	2	Identify and collect the required input data. The data may need to undergo a preparation stage, as indicated in Chapter 4
	3	Select the modelling technique(s) to be used that fit to the model's objective
	4	Select the most appropriate software tool(s) that could provide a solution to the model developed
	5	Develop the conceptual model
	6	Verify and validate the model developed
Analysis stage	1	Solve the model and extract the solution provided
	2	Analyse the solution provided and compare this with the intended objective of the conceptual model
	3	Develop alternative scenarios and compare the solution with the conceptual model's intended objectives
Implementation stage	1	Identify the most appropriate solution based on the developed scenarios
	2	Implement the model and/or the model's solution
	3	Monitor and improve the model
Post-implementation stage	1	Review the business benefits generated by the implemented model
	2	Develop a plan for reviewing the model's objectives and the data used

need to be performed to ensure the modelling process can start. Identifying the boundaries of the system under investigation, understanding the data required for the model development and operation, and understanding and identifying the most appropriate performance measures form the key elements of this stage. Following this, the modelling stage will start by considering modelling techniques to be used, software to be employed and IT systems to be used. The verification and validation of the model developed will be followed by the model's analysis stage, where the solutions obtained are to be evaluated and alternative scenarios to be employed. A valid model and reliable solution are ready now for the implementation stage. Real benefits are expected after this implementation stage, where a post-implementation evaluation should not be ignored.

Summary

It can be concluded that models have advanced in recent years; however, they are still limited in their ability to integrate all functions in a supply chain system and find trade-offs between multi-objectives at different levels and points in the supply chain.

To respond on time and in full to customer demands at different levels and points in the supply chain and to cope with its dynamic nature, to reach value added and ensure reduction in the overall cost for final products is not solely the responsibility of a single organization as it is the accumulated effort of organizations that form part of the entire supply chain. Therefore, the argument that is sustained is that modelling a supply chain from a focal company point of view cannot be effective unless its effect on other links in the chain can be evaluated. This assumes that members in the supply chain are aware of their links to other individual members in the chain.

Organizations within the supply chain will set up their objectives from operational to strategic, and each individual organization will also set their objectives to meet environmental requirements. However, no one organization in the supply chain can be responsible for setting the objectives for the entire supply chain system. Modelling a supply chain system requires working with individual organizational objectives, where there is the natural limitation of setting up the overall supply chain's system objectives. Having this type of constraint puts limitations on selecting the most appropriate modelling technique to be used and the most appropriate software package to be considered in each case.

The need for more theoretical examples is still there where new advances in the area of modelling are still in demand. Models derived from practical applications have their own benefits and demonstrate the applicability of the techniques used and their limitations. Theoretical models, however, are still required to allow for further understanding of the modelling paradigm of complex supply chain systems.

References

Abai, N H Z, Yahaya, J, Deraman, A, Hamdan, A R, Mansor, Z and Jusoh, Y Y (2019) Integrating business intelligence and analytics in managing public sector performance: An empirical study, *International Journal on Advanced Science, Engineering and Information Technology*, 9(1), 172–80

Beamon, B M (1998) Supply chain design and analysis: Models and methods, *International Journal of Production Economics*, 55, 281–94

Bertrand, J W M and Fransoo, J C (2002) Modelling and simulation: Operations management research methodologies using quantitative modeling, *International Journal of Operations and Production Management*, 22, 241–64

Blackhurst, J, Wu, T and O'Grady, P (2004) Network-based approach to modelling uncertainty in a supply chain, *International Journal of Production Research*, 42 (8), 1639–58

Bozarth, C C, Warsing, D P, Flynn, B B and Flynn, E J (2009) The impact of supply chain complexity on manufacturing plant performance, *Journal of Operations Management*, 27, 78–93

Brandenburg, M, Govindan, K, Sarkis, J and Seuring, S (2014) Quantitative models for sustainable supply chain management: Developments and directions, *European Journal of Operational Research*, 233, 299–312

Chilmon, B and Tipi, N S (2014) Modelling an end to end supply chain system using simulation, LRN Annual Conference and PhD Workshop 2014, 3–5 September 2015, University of Huddersfield

Choi, T Y and Krause, D R (2006) The supply base and its complexity: Implications for transaction costs, risks, responsiveness, and innovation, *Journal of Operations Management*, 24, 637–52

Choi, T Y, Dooley, K and Rungtusanatham, M (2001) Supply networks and complex adaptive systems: Control versus emergence, *Journal of Operations Management*, 19, 351–66

Chytas, P, Glykas, M and Valiris, G (2011) A proactive balanced scorecard, *International Journal of Information Management*, 31(5), 460–68

Day, J M (2014) Fostering emergent resilience: The complex adaptive supply network of disaster relief, *International Journal of Production Research*, 52(7), 1970–88

Dekkers, R (2017) *Applied Systems Theory*, 2nd edn, Springer International Publishing, Cham

Hearnshaw, E J S and Wilson, M M J (2013) A complex network approach to supply chain network theory, *International Journal of Operations and Production Management*, 33(4), 442–69

Jain, V and Benyoucef, L (2008) Managing long supply chain networks: Some emerging issues and challenges, *Journal of Manufacturing Technology Management*, 19(4), 469–96

Kabak, O and Ulengin, F (2011) Possibilistic linear-programming approach for supply chain networking decisions, *European Journal of Operational Research*, 209, 253–64

Lambert, D M and Cooper, M C (2000) Issues in supply chain management, *Industrial Marketing Management*, 29(1), 65–83, doi:10.1016/S0019-8501(99)00113-3

Manuj, I and Sahin, F (2011) A model of supply chain and supply chain decision-making complexity, *International Journal of Physical Distribution and Logistics Management*, 41(5), 511–49

Mentzer, J T, Dewitt, W, Keebler, J S, Soonhoong, M, Nix, N W, Smith, C D and Zacharia, Z G (2001) Defining supply chain management, *Journal of Business Logistics*, 22, 1–25

Milgate, M (2001) Supply chain complexity and delivery performance: An international exploratory study, *Supply Chain Management: An international journal*, 6, 106–18

Min, H and Zhou, G (2002) Supply chain modeling: Past, present and future, *Computers and Industrial Engineering*, 43, 231–49

Mitroff, I I, Betz, F, Pondy, L R and Sagasti, F (1974) On managing science in the systems age: Two schemas for the study of science as a whole systems phenomenon, *Interfaces*, 4, 46–58

Morris, W T (1967) On the art of modeling, *Management Science*, 13, B707–B717

Pathak, S D, Day, J M, Nair, A, Sawaya, W J and Kristal, M M (2007) Complexity and adaptivity in supply networks: Building supply network theory using a complex adaptive systems perspective, *Decision Sciences*, 38(4), 547–80

Pidd, M (2010) Why modelling and model use matter, *Journal of the Operational Research Society*, 61, 14–24

Riddalls, C E, Bennett, S and Tipi, N S (2000) Modelling the dynamics of supply chains, *International Journal of Systems Science*, 31, 969–76

Serdarasan, S (2013) A review of supply chain complexity drivers, *Computers and Industrial Engineering*, 66(3), 533–40

Shapiro, J F (2001) *Modeling the Supply Chain*, Thomson Brooks/Cole, Australia

Skyttner, L (1996) *General System Theory: An introduction*, Antony Rowe Ltd, Chippenham

Slats, P A, Bhola, B, Evers, J J M and Dijkhuizen, G (1995) Logistic chain modelling, *European Journal of Operational Research*, 87, 1–20

Turner, N, Aitken, J and Bozarth, C (2018) A framework for understanding managerial responses to supply chain complexity, *International Journal of Operations and Production Management*, 38(6), 1433–66, doi:10.1108/IJOPM-01-2017-0062

Weaver, W (1948) Science and complexity, *American Scientist*, 36, 536–44

Wilding, R (1998) The supply chain complexity triangle: Uncertainty generation in the supply chain, *International Journal of Physical Distribution and Logistics Management*, 28(8), 599–616

Data in the supply chain

03

LEARNING OBJECTIVES

- Understand data and be able to differentiate between and define small and big data.
- Enquire and reflect on questions related to data in the supply chain.
- Understand the difference between small and big data in different cases in the supply chain.
- Evaluate some of the challenges in working with data.

Introduction

Organizations have the opportunity to access large sets of data, such as product characteristics, inventory positions, demand, forecasted data and many more, where the effort to acquire data has become easier over the years. This is due to the advances in technology and the availability of different IT tools that have the power to save large sets of data, benefit from a user-friendly interface and provide a large array of analyses. The availability of new tools such as sensors and RFID, and their ability to connect in real time with devices that allow for data manipulation, have given organizations a new avenue for collecting data that was not possible previously. The most relevant data to the supply chain needs to firstly be identified, then extracted in the required format and finally prepared to be ready for the analysis (later on in this chapter, a data analytics methodology will be discussed). Due to the large sets of data available to us, understanding and making use of the data could be an overwhelming task.

The challenge to identify, obtain and work only with the most relevant set of data that forms part of the decision-making process is still present. However, to this can be added the challenge of large amounts of data that comes in different forms, formats, at different time intervals, and through different systems. Data that is left without being analysed will lose its value and in a short amount of time it could become obsolete. Still, this is not to be confused with historical data, which is key in record keeping, developments and innovations. Fosso Wamba et al (2018) argue that what is not yet clear is how big data affects operations and supply chain management processes. Big data is generated within an organization; however, a large amount of data that is having a direct impact on an organization is also generated outside the organization. Understanding, adopting, developing intelligent tools and approaches that access, manipulate, maintain and analyse these types of data is what organizations are seeking to do in order to maintain a competitive edge.

An analyst will be keen to understand what data is required in a supply chain system to allow for modelling and analysis of various processes and performances within the supply chain. At the same time, large sets of data are generated and are available within an organization as well as external to an organization that relate directly to various aspects of an organization. An understanding of what type of data is available and how this data will be required, manipulated, operated and analysed is paramount within a supply chain context.

Within an organization that forms part of the supply chain, data is accumulated from an array of functions and a range of different initiatives that take place, such as: organizational data, master data, sourcing data, production planning type data, process type data, sales and demand data, distribution type data, storage and location information, financial, human capital, asset management, customer service, quality management type data, and a range of business process analysis type data. This list is then split into a number of different smaller sections, each with their importance and relevance. It may appear that this classification could go on; however, all of these details would need to be captured within an information system and evaluated to form part of value creation for an organization.

To the internally generated data, we are now adding external data. External data may have pre-established, known characteristics; however, a number of atypical types of data are being generated and influence, to some extent, the decision-making within the supply chain system. External data may not necessarily be generated in a structured format. External data could come from sources such as social media, video and audio recordings, newspapers, photographs and others, where their format is not exactly controlled by a particular goal of an organization, or they are not given in a predefined format.

Questions that researchers and practitioners ask regarding data are listed in Table 3.1. Some very interesting questions from previously published research are compiled in Table 3.2.

Table 3.1 Sample of data in the supply chain questions

Questions	Relevant to
What data is currently collected within the organization? What data quality tests are carried out internally to evaluate data quality? Where is the data currently saved? Which team is responsible for data management and data quality? What is the data flow within an organization? Have the roles that are authorized to change and manipulate data been clearly defined within each department in an organization? How often is the data saved and backed up? For how long is the data kept? What is the cost associated with collecting and storing data within an organization? What is the cost associated with maintaining data quality?	Internal to an organization: everyone working with data, managers responsible for data management and reporting, IT managers Researchers: in the area of operational management, strategic management, logistics and supply chain management
What external data is currently generated regarding products or services of an organization? Who is responsible for analysing external data? What knowledge has been generated as a result of external data? What data validity tests are carried out to external data to ensure data quality? How much does it cost to analyse external data? What technology needs to be put in place to analyse external data? Is the external data open access? Who else has access to this data?	Internal to an organization: strategic managers, operational managers, process managers, data warehouse managers, IT managers External to an organization: for example, retailers selling a product and collecting data about this; service organizations such as port operators, transport organizations, railway companies

(continued)

Table 3.1 (Continued)

Questions	Relevant to
Do current performance measurement systems within an organization capture data related measures? What are the data performance measures that refer to aspects of data cost, data reliability, data quality and data accessibility?	Internal to an organization: IT managers Researchers: in the area of operational management, performance measurement, logistics and supply chain management

Table 3.2 Big data questions identified in the literature

Questions	Sources
How does big data impact supply chains? What are the key success factors for big data implementation in supply chain management and logistics? What theory best explains the potential long-term value big data can bring to supply chain partners?	Richey et al (2016), p 712
How can firm performance be improved using big data analytics capability and business strategy alignment?	Akter et al (2016), p 113
What different types of big data challenges are theorized/proposed/confronted by organizations? What different type of big data analytics methods are theorized/proposed/employed to overcome big data challenge?	Sivarajah et al (2017), p 263
In what area of supply chain management is big data analytics being applied? At what level of analytics is big data analytics used in these supply chain management areas? What types of big data analytics models are used in supply chain management? What big data analytics techniques are employed to develop these models?	Nguyen et al (2018), p 255
What is the influence of big data analytics personnel expertise capabilities on supply chain agility? What is the interrelationship of big data analytics personnel's expertise capabilities?	Mandal (2018), p 1202

(continued)

Table 3.2 (Continued)

Questions	Sources
What are the effects of big data and predictive analytics on social performance and environmental performance? How do human (technical and managerial) skills and big data culture (organizational learning and data-driven decision-making) help to build big data predictive analytics? What are the effects of organizational culture on the relationship between big data predictive analytics and social/environmental sustainability?	Dubey et al (2019), p 535
How do big data and analytics methods coalesce to provide actionable insight? How do resource bundling and orchestration practices influence organizations' big data and business analytics leveraging capacity? What organizational capabilities can big data analytics enable or automate, and what is the effect of doing so?	Mikalef et al (2020), p 2

A data analytics methodology

For organizations forming part of the supply chain to gain the full benefit of operating with data, they may consider employing a data analytics methodology (Gupta, 2018). Other authors have also presented data infrastructure stages (see Zhan and Tan, 2020). Within this study the following framework is developed, where the need for individual steps is explained below and captured in Figure 3.1.

Step 1: Data identification

This can be considered as the initial step in the process as time and resources should be allocated to identify what potential data may be available for the task at hand. At this step, knowledge of what constitutes the data to be collected is important, as well as what data is available.

The sources from where the data needs to be collected need to be identified at this step. Data may or may not be available, and particular managerial tasks need to be put in place for identifying the data sources. There are situations, where there may be the possibility for a particular aspect of data to be collected; however, the organization may not have collected this type of data, or they may have collected it but not necessarily put in the performance measurement system. There are other situations where large sets of data are being generated but they do not end up into any measurement system. Identifying the required data may be a lengthy process, but cannot be ignored as it may affect the next steps in the data analysis process.

Step 2: Data collection

This is a very important process that needs to take place to ensure that the outcomes of the analysis have the potential to generate value to the organization and to the supply chain. Particular methods have been used by researchers and practitioners to collect data (via questionnaires, surveys, generated by different systems or simulations and saved within a data warehouse system and others). Each individual method used to collect data will require a process of verification and validation applied to the data collection instrument used. An analyst needs to ensure that the data they want to collect is the actual data intended for collection, and the instrument used is free of any errors. Again, it is relevant to stress that time must be invested in identifying the best instrument or electronic system to be used, and verifying and validating this before the actual process of collecting data takes place.

Step 3: Data preparation

At this step, data is brought into the format to be evaluated. Data cleaning takes place at this stage, where the input values are checked for errors or missing values, and duplicates in data are deleted. At this step, the outliers in data are identified (this may be by using simple statistical techniques, such as averages) and these may be eliminated or smoothed out (readjusted using calculations such as regression models). Checks may also identify data that has odd values, values that may have been entered by error, and at this step this can also be corrected.

Step 4: Data analysis

As soon as the data is clearly understood, and has been prepared in the correct format ready for analysis, the process of analysing the data can start. Depending on the data type and format, the correct method(s) needs to be employed to analyse the data.

Step 5: Results validation and verification

After the data has been analysed, Gupta (2018) proposes as a following step *model building*. At this stage in the process, data may present different patterns, or characteristics that will allow an analyst to construct models (for example statistical models, such as regression equations, or simultaneous equations, or particular distributions can be identified and used to generate more data, and so on).

Step 6: Implementation of results

Using models, results will be generated, therefore this step is concerned with the *results* obtained after applying particular models. The models as well as the results will need to go through a process of validation and verification of results.

As soon as all the results have been verified, and they present potential benefits to the organization, they can now be implemented in practices. Therefore, this last stage in the methodology is concerned with data *implementation*.

Zhan and Tan (2020) present five stages in their framework. Stage 1 considers data capture and management. Stage 2 is comprised of data cleaning and interrogation. These aspects are captured within the data preparation in this case. Stage 3 in their evaluation is formed of data analytics. Stage 4 considers competence set analysis. Stage 5 is built around information interpretation and decision-making.It is also considered relevant in this case that aspects of implementations should be taken into consideration.

Figure 3.1 A data analytics framework

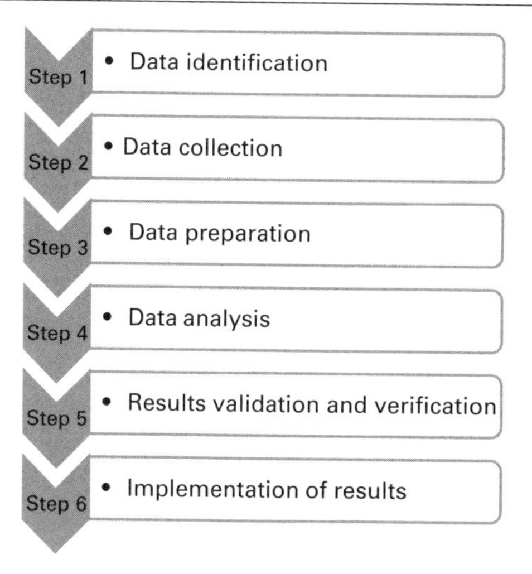

Data type

In broad terms, data can be differentiated in two types: quantitative and qualitative. However, we may also have mixed data that has a numerical as well as a text format.

Quantitative data has a numerical format. Quantitative data can be further split into, for example, discrete, continuous and binary data. *Discrete* data can be counted and in general has an integer value (for example: 1, 2, 3, ...). This type of data is also referred to as *categorical* data. *Continuous* data is data that can be measured on a continuous scale and can have any format (for example, length, weight, temperature, and so on). This type of data is known as continuous as between any two values of this data it can be considered that there are an infinite number of other values (Kubiak and Benbow, 2016). When evaluating processes in real time, continuous-type data can be collected. This is considered more accurate, as it gives the analyst the opportunity to observe any small variations between values, or between discrete values. However, continuous data tends to generate big volumes, and in some instances this aspect may be avoided, depending on the details required for the analysis. *Binary data* in a numerical format is data that has only two values, such as 0 or 1. Binary data in the format of 0/1 is also the data that is used in computer language. One other data category that can be recognized here is *interval data* and this is measured in the form of intervals. Quantitative data can also be expressed in *ratios*. Both ratio and interval data are continuous data types.

Qualitative data is data that is in the format of text or a story. Qualitative methods need to be used to collect and analyse this type of data. A range of methods are used by researchers and practitioners to collect qualitative data, such as interviews, questionnaires, diary entries and others. Data of a qualitative nature can also be observed in published case studies, summary reports, consumer feedback, and so on. A number of methods have been developed to analyse qualitative data, such as Delphi techniques, content analysis, thematic analysis and others. Different types can be attributed to the qualitative data, such as *binary* format where only two numerical values are registered (for example yes/no, or true/false); or data can be categorized as *nominal* where different categories can be attributed to it; however, there is no specific order of rank to the data. Qualitative data is categorized as *ordinal* where an order is specified. Both nominal and ordinal data are discrete in nature.

Mixed data is data that considers numerical data, text data and any other type of data. Other type of data could be in the form of video or voice/audio, and they many need to be converted into a format that can be analysed. Different methods need to be considered in this case, where methods to analyse this type of data can be predefined, before the data has been collected, or need to be established as soon as the analysis is faced with the data received, obtained, or extracted by the system.

A number of algorithm and computer programmes are very specific on the data they can handle.

Small and big data in the supply chain

Data is the omnipresent force in all areas of the supply chain, from the movement of goods, to planning, to manufacturing, to forecasting, to the financial transaction, to services, to knowledge and information generation, to name a few of the activities. The use of reliable, accurate data has the opportunity to facilitate increased visibility in the supply chain, to allow for transparency, coordination and better communication. Supply chain analytics uses data from internal and external sources in different formats to enhance decisions, facilitate knowledge and provide up-to-date information to logistics and supply chain managers. The majority of activities and operations that need to be performed in the supply chain require data, and going further require reliable data.

Data and big data are used in the supply chain to facilitate analytics such as reporting, quality control, inventory control, forecasting, production planning, distribution planning, scheduling, real-time vehicle routing, capacity planning, facility location, simulation and optimization.

There are different levels of interpretation when it comes to what can be defined as small and big data. This section aims to provide further clarity to the terminology used.

Small data

As we see, we exist in a pool of large volumes of data. However, small data is also a phenomenon that is present in our day-to-day operations. While big data is generated at an incredible speed being pushed forward by technological and societal forces (Faraway and Augustin, 2018), at the same time, it can be recognized that a larger number of small data sets are also generated. Small data tends to be required to answer a particular question, or attend to one specific goal, or be part of a dedicated project (Berman, 2018).

Small data in general could be owned by a research group within a particular institution, or it can be data collected for one specific project. This data may only serve the purpose of a particular aspect of a project or a particular project and will be saved in a particular file or files, following a set format with a clearly defined structure. One other characteristic of small data is that it is probably saved only for the duration of a particular project and then deleted, serving the purpose of that particular project alone. For example, in higher education, we recognize a large number of datasets with small data saved for individual students' projects, where data is collected to serve the purpose of one particular project and it is deleted after a short and specified duration of time. Data in this case may have different formats of quantitative or qualitative structure and the file formats may be of different sizes, but they could still be classed as small data sets. A large number of small datasets may collectively result in large volumes of data, and these may be regarded by the analyst as big data due to its volume and complexity of handling.

However, as the explosion of big data approaches, it is also relevant to highlight that there is incredible power in small data. There are instances where the answer to a particular question can be fully satisfied by small data without the need to make use of big data or without the requirements of big datasets. A large number of projects in academic and industrial settings are designed to only use small data to obtain the answer to the set project objectives.

For a number of years, small data has governed the research arena, and it will continue to be valued and used.

Big data

Big data, as the names suggests, is the data that goes beyond the characteristics presented above of small data. Small data has a clearly defined size, structure and purpose, whereas big data is characterized by very large datasets with complex characteristics, varied structure and form, and is generated and stores in a combination of structured and unstructured formats, without necessarily a clearly defined goal (see Figure 3.2).

The prospect of working with and evaluating big data has attracted the interest of a large number of researchers, each putting forward definitions of big data and providing different forms to characterize it. A number of studies have provided empirical evidence of the benefits of big data such as in the areas of innovation, agility and competitive performance (Mikalef et al, 2020). To distinguish big data from other forms of data, researchers have

Figure 3.2 Big data characteristics

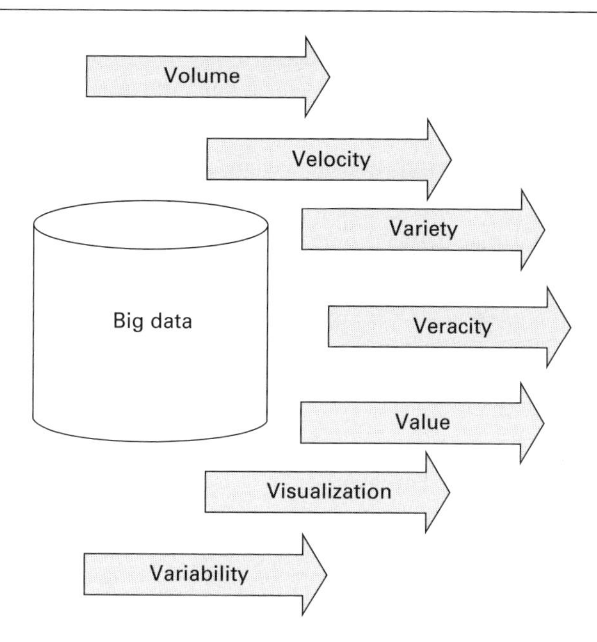

discussed dimensions such as volume, variety and velocity (Williams, 2016; Berman, 2018). Richey et al (2016) refer to four dimensions of data, namely volume, velocity, variety and veracity. Addo-Tenkorang and Helo (2016) add to these four the dimension of value.

- **Volume** refers to the amount or size of data and databases that are available, or data that is generated and requires storage (George et al, 2016). Different social media, the internet, other devices such as phones, sensors and so on, are in a position to generate new data in different forms and formats. New platforms have emerged that allow these kinds of data to be stored and manipulated.
- **Velocity** represents the speed at which the data is being generated and made available. A number of devices are set to generate data and this is being done at an incredible speed (George et al, 2016). This may be from different set devices such as controllers, to agents that are set to indicate when, for example, inventory has reached a certain level, simulations are also set to generate data and many others. However, a number of questions still remain: Will all generated data be saved? Or used? Will it have the potential to generate value?

- **Variety** is the type of data that is being created in structured, semi-structured or unstructured format. Data in varied forms has been used over a number of years, and there is an immense volume of varied types of data. Within a supply chain, data is being collected from a number of different sources (such as point of sale data, membership cards, RFID tags, GPS locations, barcodes, etc), each coming with a different data format. At the same time, organizations recognize more and more the value hidden in unstructured data, where new analytical techniques are being adopted to provide analysis of data of unstructured format.
- **Veracity** refers to the way in which the correctness, accuracy and quality of data is verified. Taking this step confirms the reliability of the data (Ittmann, 2015).
- **Value** is the added worth or usefulness data can generate as a result of the information provided. The availability of data alone is not sufficient if this can't be turned into value.

Other authors working with dimensions of data are bringing forward notions such as visualization and variety (Sivarajah et al, 2017).

- **Visualization** refers to the way in which data is represented, to make better use of what the data represents, to enhance the decision-making process. Easy accessibility and readability of data allows for better decision-making.
- **Variability** in data indicates that data's meaning may change due to analysis or over a short period of time. To understand variability in data, Sivarajah et al (2017) mention the use of sentiment analysis.

Some of the differences between big data and small data as adapted from Berman (2018) and their application in a supply chain environment are depicted in Table 3.3. They have been grouped by characteristics such as aim/goal of the data, its location, structure, preparation, longevity, analysis (Berman, 2018). However, the list could be extended with the addition of other characteristics such as quality, validation, security, and others. The intention here is to provide an understanding the difference in operation with small and big data, as moving through differed categories of data characteristics it becomes apparent that operating within a supply chain setting there are many cases when the data that is being considered is 'big data'.

Table 3.3 Small and big data in the supply chain

General characteristics	Big data (BD) vs small data (SD)	Supply chain implications
Aim/goal of data	BD – flexible goal	BD – examples could be consumer feedback on social media of the characteristics of a particular product; sensor data collected from a manufacturing machine in relation to product quality
	SD – specific goal	SD – used to provide daily reports, daily transactions, used for forecasting, analysis of strategic nature, analysis of an operational nature
Location of data	BD – online; saved on internet, cloud	BD – different internet servers in different counties, not necessarily associated with a particular supply chain
	SD – internally saved on a particular personal computer, or within a department, or in the ERP system of an organization	SD – contained within one organization in the supply chain, and may not be shared with other members in the chain; or a particular password protected space on cloud related to one aspect that is shared with all members of the supply chain
Data structure	BD – contains a combination of structured, semi-structured and unstructured format of data. This data could be internal and external to the supply chain	BD – unstructured format of email communications, reports in printed and electronic format, diagrams, maps, sensor data, RFID data and so on, and data in structured format, texts, tweets, data in numerical and non-numerical format
	SD – highly structured data	SD – data in structured format from systems such as MRP, MRP II, ERP, CRM, EPOS and warehouse management system

(*continued*)

Table 3.3 (Continued)

General characteristics	Big data (BD) vs small data (SD)	Supply chain implications
Data preparation	BD – generated by many sources and prepared by many involved SD – individual users within an organization or a department	BD – prepared by many members of a particular supply chain, or a competing supply chain, or external to the supply chain SD – prepared by users responsible for data input, data analysis within a department of an organization in the supply chain, or the central team of an organization's supply chain that shares data within the same organization or their identified partner organizations
Data longevity	BD – data may be required to be stored permanently (Berman, 2018) or for a longer period of time SD – data may be saved for a limited time period for the duration of a project, or saved for a longer period of time that still may be specified	BD – for example, emails regarding particular characteristics of a product; or data related to a particular test or analysis that is relevant for the development of a new product, system or process in the supply chain SD – for example, data related to implementation projects within an organization is to be saved for the duration of the project; data used for the development of a new product may need to be saved for the duration of production only, or for a particular time period after this
Data analysis	BD – data is analysed using different tools and at different stages SD – data collected for a particular task or project is analysed within that project	BD – data required for analysis may be in a different system, in different locations, and the analysis requires identification of data, cleaning of data, formatting, analysis and re-analysis SD – product development projects, optimization routing projects; distribution planning projects; product innovation project and so on

Challenges in working with data

Big data has the potential to present a number of advantages to organizations when managed, processed, stored and analysed in line with its intended purpose. It is relevant to highlight here that the real value of data in its big and small format is not necessarily in the 'data' itself, but rather it is in the way this data is managed and evaluated, and how the value from the data is extracted and implemented. Big data opens the door for 'big prospects' as it has the potential to generate new knowledge, provide new insights, and to enhance the decision-making process. However, it also leads to a number of challenges.

Some of the challenges discussed in the following section include dealing with large volumes of data, the management of different type of data, understanding data quality, the complexities linked to the integration of data through different systems in the supply chain, and challenges linked to infrastructure and data warehouse architecture.

Sivarajah et al (2017) indicate that data challenges can be grouped in three distinctive categories: characteristics of data challenge, data process challenge (which is linked to the way in which data is identified, extracted, saved, processed, analysed and displayed) and data management challenge (which may include aspects of security, accessibility rights, data governance, privacy, ethics) (Sivarajah et al, 2017, p 265). Some data challenges are detailed in the section below.

The management of a large volume of data

The availability of large sets of data offers a number of advantages. However, this may also generate additional cost in extracting and analysing the data. This may generate confusion if data appears to be aggregated, or is presented in different forms that will bring uncertainty in taking a decision. Data available in large volumes, saved in different systems in different data formats, may be missed in the analysis due to researchers and analysts being unaware where all these data resides. It may also be the case that only samples of large volumes of data are being analysed, where assumptions will be considered for the rest of the data.

Larger volumes of data must be saved and maintained over a period of time. Data generated internally by an organization becomes the responsibility of the organization where data will be saved and managed internally using different platforms. However, data is also generated externally and may comprise details about a particular organization, although this data may be lost as it does not belong to the organization the external data may generate value to. The volume of externally generated data may not always be predictable; therefore, an organization is not always in a position to plan for the resources required to analyse, save and maintained externally generated data.

For large volumes of data, an organization needs to have the computational power to be in a position to access this data when required and to carry out analysis at the time and frequency these are required. Some companies make use of *parallel processing* to increase data processing speed, and others make use of the *grid computing approach* (Somani and Deka 2017, p 41).

Researchers have brought forward arguments in favour of 'big data' when small data is not presenting sufficient details in trend spotting (see for example Irani et al, 2018), and therefore more data is to be used for analysis. However, a large volume of data generated on a continuous form and at different time intervals may never be analysed in its totality (Berman, 2018), although this data will be generated and stored, making use of current resources.

The example presented in Figure 3.3 aims to indicate how data collected over a one-year period may provide a different forecasted answer when analysed, as opposed to using a larger volume of data for analysis. In this particular example, if previous events are known not to repeat in the current year, analysing a larger volume of data does not add value, as the resulting forecasted value will not be the expected value; however, if it is known that past events may repeat in the future, a larger volume of data should be analysed.

The challenge of working with different types of data

Data has different forms: structured and unstructured; internal and external. When data is structured, clear categories are indicated, access to this data is predefined, the storage space is indicated where manipulating structured data appears more manageable. Big data and small data can be characterized by structured data. Examples of structured data could be transactional data that is saved and can be extracted from a machine requirements planning (MRP) system, from a distribution requirements planning system (DRP), ERP system, etc, or data saved as barcode, demand forecast data, sales data, financial data on purchase cost, transportation cost, production cost, manufacturing cost and so on. Berman (2018) points out that nearly all data in its raw form is unusable and unstructured. Marr (2016) talks about the limitation of structured data in the sense that this could provide a limited view of the

Figure 3.3 Example of data recorded over one-year versus five-year period

Value

Time (1 year of data)

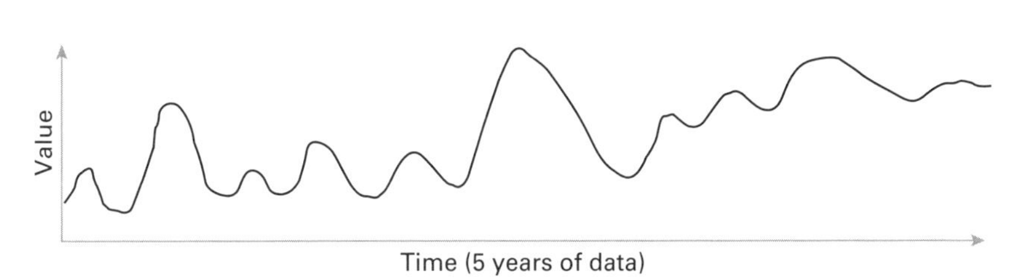

whole story of what data could tell, is only a small portion of the whole data, and is less rich than unstructured data. However this may be the case, unstructured data generated without a predefined goal for analysis brings the challenge of extracting information and knowledge that will add value to an organization without taking an extensive amount of time for analysis and incurring unnecessary cost. Unstructured data in its many formats, such as documents, web logs, call logs, sensor data, text messages, email records, tweets, audio or video files, photographs, images and machine generated data has the potential to bring new characteristics and dimensions regarding data that has not yet been captured in a structured format.

Understanding data quality

One of the key challenges in working with data is data quality. Data that does not reach the required data quality parameters will have little or no relevance in achieving added value as a result of data analysis, regardless of the analytical technique used for investigation. Some of the parameters that are linked to the quality of data are accuracy, consistency, completeness and timeliness of the data (Hazen et al, 2014).

Accurate data refers to the fact that the data represents what it is expected to represent. For example, an organization may save data about a product in two different systems. In this case, the data available and saved about one particular product should be the same in both systems. Other examples could be the address saved about one of the company's suppliers. If the address saved represents exactly the location where the supplier is situated, this implies that the record saved is an accurate representation, and is classed as accurate data. However, if this record is saved in more than one place, and if for example the supplier's address has changed, this change needs to be reflected in all saved locations of this data. To ensure data accuracy within a system, within an organization, tests need to be put in place at regular time intervals to verify data accuracy. Checking data accuracy can be a very elaborate and time-consuming task, therefore appropriate time needs to be considered when managing and verifying the accuracy of the data saved and available to an organization.

Data consistency is seen in relation to the *format* and *structure* attributed to the data (Hazen, et al, 2014). For example, a customer's address that is recorded in two systems used by an organization, such as an enterprise resource planning system and a data warehouse management system, must use the same fields for the address (name, street name, house number, town/city, postcode country) in the same format. Data consistency is particularly relevant when data needs to be transferred from one system to another, and required to compile reports.

Complete data refers to records for which all values have been collected and there is no missing data. For example, the sourcing department wants to place an order for a number of raw materials that form part of a product. Incomplete details on the bill of material records results in errors in placing an order with suppliers, which will then generate inconsistencies and delays in manufacturing this product. One other example could be the records saved about a delivery order. Any missing field in saving all the required records in a delivery order could result in a missed delivery, and consequently add unnecessary cost to delivery and cause delays. To ensure completeness of the data saved, a number of software packages could have compulsory data fields set up. Another measure that could be put in place to ensure that the data saved is in a complete format is to run spot check reports for a particular data field to ensure data completeness.

A number of dimensions are attributed to the timeliness of data. One of these refers to the fact that the data saved in the system is current and up to date. For example, to be able to calculate the reorder position for placing an order for a particular raw material, current inventory data is expected to be saved in the system. One other example could be that calculating the cost of a product requires up-to-date exchange rate information when the product is sold internationally. A time-related dimension attributed to data is the frequency of collecting and saving the data in the system.

Data infrastructure and data warehouse

One of the biggest challenges when dealing with big data is the cost of the infrastructure required to save and maintain the hardware equipment or the cloud computing technologies. Organizations within a supply chain will have different facilities, different access to resources, and different infrastructure, with internal users having different level of access to data and knowledge to operate and analyse data. Therefore, the established IT links between different functions both within an organization and externally are particularly important when dealing with data-related issues.

Big data benefits from advanced technologies that enable its analysis (NoSQL Databases, Big Query, MapReduce, Hadoop platform, WibiData, Skytree, EDW 3.0 Enterprise Data Warehouse (EDW 3.0)) (Sivarajah et al, 2017; Morton, 2014). Data visualization software packages such as Tableau offer a number of visualization functionalities and smart dashboards. Also, software packages such as IBM SPSS Statistics and SPSS Modeler allow for a number of data functionalities from data cleaning, data preparation to predictive analysis and data modelling that are supported by advanced visualization tools.

Data collection points in the supply chain

There are a number of different points in the supply chain where data can be collected. However, individual organizations, as part of the supply chain, may only have access to data owned by them. Still, the data collection points within an organization play a significant role in various aspect of analysis and decision-making processes, and need to be carefully understood and considered.

Data collected at the end of a process

Data collected at the end point has an aggregate view that comprises the results of activities and processes at individual points in the system. Data collected at this point may bring a holistic understanding of the individual operations. However, it has less power in giving information on where a particular problem may have occurred. Data collected at this point is in general used for yearly reporting and is also used for benchmarking. Aspects of product and process type data collected at the end of a process are in general shared across the supply chain if they do not present any security or confidential details.

Data collected at the start of a process

It is important to set data collection points at the beginning of a process to give managers an understanding of the entry values of different entities, for example the inventory position of raw materials, temperature before starting a particular operation, volume, quantity, density, and so on. Collecting data before and after a process can bring details of the efficiency of the particular process being analysed. In the supply chain, the entry point of a process can be the end point of a previous process or previous processes.

Data collected in progress

Collecting data in progress at different points in the process may allow organizations to take corrective action earlier on in the process. This task should, however, be considered with caution as it will have the potential to generate large volumes of data at multiple points in the process, data that will require storage and time invested for analysis. There are new approaches considered nowadays, where automated processes are put in place for data to be collected and analysed this way, and corrective actions to be considered at the same time. Artificial intelligence, intelligent agents and machine learning are some of the tools that are or can be used for this type of tasks.

Data collected at divergence and convergence points in the process

Data collected at a divergence point is important as it gives details about what is used as input data in the process, and what is particularly split to follow different paths, especially when the split is not equal and the process follows particular process decisions. As well as the divergence points, collecting data at convergence points in the process gives an understanding of what input data goes in the process from different entry points (Kubiak and Benbow, 2016).

Data collected between different processes, functions and departments within an organization

For example, products are being manufactured and are now waiting in the local warehouse to be packed ready for delivery. Different departments are responsible for different activities and they may set data points for their individual activities within their own department. However, data collection points should be set at transit points between different departments. Not all departmental data that has been collected and analysed for a product or set of products is relevant to other departments in the chain, and therefore consideration needs to be given to what data and data sets needs to be transferred between departments.

Data collected and transferred outside an organization

As we have seen from the previous discussion, data in the supply chain is generated outside an organization and this is being generate from social media, to many Internet of Things (IoT) and Internet of Services (IoS) devices. An organization may have limited access to some of the external devices and points of data collection; however, partners within the same supply chain may agree to share and collect data outside their organization. There are a number of implications when data is being generated outside an organization, but there are also a number of benefits when data is shared and other organizations along the chain are in a position to benefit from some of the shared data, such as inventory positions, forecasted data, delivery information, and so on. Data collection points outside an organization may be established as soon as a product or service has left the organization, is in transit, has reached the end of the transportation, and before data enters an organization.

Data recorded on the reverse cycle of the supply chain

The reverse movement of products in the supply chain is created by returned products, recycled, reused, remanufactured and repurposed products or parts/modules of products. Data collection is key in the returned function in the supply chain. However, this depends very much on the returned activity and the industry involved. Data collection points can be established not only inside an organization, indicating the opportunities for recycling, reusing and repurposing, but also outside an organization at different points in the chain. The data collected in this scenario has more dimensions than the position of the collection point in the chain. This data should also carry an indication on the volume, quantity returned, time of return and quality of the product returned.

Summary

Working with data is not new, but it is becoming more and more important for us to understand data and its importance within and outside an organization. This chapter aimed to explain what data is, and the different types and characteristics of data. Aspects of small and big data, structured and unstructured, quantitative and non-quantitative data, formatted and/or unformatted data have been considered. Definitions of small and big data have been given, and they have also been compared through different categories relevant to the supply chain. There are a number of aspects that are not clearly envisaged at the start of a project regarding data, where externally generated data may bring further understanding to the issues being investigated. Some of the challenges in working with data in the context of supply chain have been discussed in this chapter. May other aspects of data can be explored, but it is important to understand the magnitude of issues that could arise if data is not available or it has not been considered for analysis at different key points in the system of supply chain. However, to be in a position to understand these aspects of data, a number of questions have been asked and explored in this chapter to help practitioners and researchers further develop their level of enquiry into their individual projects.

References

Addo-Tenkorang, R and Helo, P T (2016) Big data applications in operations/supply-chain management: A literature review, *Computers and Industrial Engineering*, 101, 528–43

Akter, S, Wamba, S F, Gunasekaran, A, Dubey, R and Childe, S J (2016) How to improve firm performance using big data analytics capability and business strategy alignment? *International Journal of Production Economics*, 182, 113–31

Berman, J J (2018) *Principles and Practice of Big Data: Preparing, sharing, and analyzing complex information*, 2nd edn, Elsevier Science and Technology Books, London and Burlington

Dubey, R, Gunasekaran, A, Childe, S J, Papadopoulos, T, Luo, Z, Wamba, S F and Roubaud, D (2019) Can big data and predictive analytics improve social and environmental sustainability? *Technological Forecasting and Social Change*, 144, 534–45, doi:10.1016/j.techfore.2017.06.020

Faraway, J J and Augustin, N H (2018) When small data beats big data, *Statistics and Probability Letters*, 136, 142–45, doi:10.1016/j.spl.2018.02.031

Fosso Wamba, S, Gunasekaran, A, Dubey, R and Ngai, E W T (2018) Big data analytics in operations and supply chain management, *Annals of Operations Research*, 270(1–2), 1–4, doi:10.1007/s10479-018-3024-7

George, G, Osinga, E C, Lavie, D and Scott, B A (2016) Big data and data science methods for management research, *Academy of Management Journal*, 59(5), 1493–507

Gupta, D (2018) *Applied Analytics Through Case Studies using SAS and R: Implementing predictive models and machine learning techniques*, 1st edn, Apress, Berkeley, CA doi:10.1007/978-1-4842-3525-6

Hazen, B T, Boone, C A, Ezell, J D and Jones-Farmer, L A (2014) Data quality for data science, predictive analytics, and big data in supply chain management:

An introduction to the problem and suggestions for research and applications, *International Journal of Production Economics*, 154, 72–80
Irani, Z, Sharif, A M, Lee, H, Aktas, E, Topaloğlu, Z, van't Wout, T and Huda, S (2018) Managing food security through food waste and loss: Small data to big data, *Computers and Operations Research*, 98, 367–83
Ittmann, H W (2015) The impact of big data and business analytics on supply chain management, *Journal of Transport and Supply Chain Management*, 9, e1–e9
Kubiak, T M and Benbow, D W (2016) *The Certified Six Sigma Black Belt Handbook*, 3rd edn, ASQ Quality Press, Milwaukee
Mandal, S (2018) An examination of the importance of big data analytics in supply chain agility development, *Management Research Review*, 41, 1201–19
Marr, B (2016) *Big Data for Small Business for Dummies*, John Wiley & Sons, Chichester
Mikalef, P, Pappas, I O, Krogstie, J and Pavlou, P A (2020) Big data and business analytics: A research agenda for realizing business value, *Information and Management*, 57(1), doi:10.1016/j.im.2019.103237
Morton, J (2014) *Big Data: Opportunities and challenges*, BCS, The Chartered Institute for IT, Swindon
Nguyen, T, Zhou, L, Spiegler, V, Ieromonachou, P and Lin, Y (2018) Big data analytics in supply chain management: A state-of-the-art literature review, *Computers and Operations Research*, 98, 254–64
Richey, R G, Morgan, T R, Lindsey-Hall, K and Adams, F G (2016) A global exploration of big data in the supply chain, *International Journal of Physical Distribution and Logistics Management*, 46, 710–39
Sivarajah, U, Kamal, M M, Irani, Z and Weerakkody, V (2017) Critical analysis of big data challenges and analytical methods, *Journal of Business Research*, 70, 263–86
Somani, A K and Deka, G C (2017) *Big Data Analytics: Tools and technology for effective planning*, CRC Press, Boca Raton, FL
Williams, S (2016) *Business Intelligence Strategy and Big Data Analytics: A general management perspective*, Morgan Kaufmann, Amsterdam
Zhan, Y Z and Tan, K H (2020) An analytic infrastructure for harvesting big data to enhance supply chain performance, *European Journal of Operational Research*, 281(3), 559–74

Bibliography

EMC Education Services (2015) *Data Science and Big Data Analytics: Discovering, analyzing, visualizing and presenting data*, Wiley, Indianapolis, IN
Gupta, B (2016) *Interview Questions in Business Analytics*, Apress, Berkeley, CA
Sahay, A (2017) *Data Visualization: Volume II – Uncovering the hidden pattern in data using basic and new quality tools*, Business Expert Press, New York

Supply chain performance measurement systems

04

LEARNING OBJECTIVES

- Understand performance measures and how they are set.
- Ask and reflect on questions related to performance measures in the supply chain.
- Understand the steps required to develop a performance measurement system suitable for analysis in a supply chain setting.
- Evaluate the role of data required to operate a reliable performance measurement system.

Introduction

Performance measures are set and used within an organization in order to monitor, control, test, evaluate, reflect, improve, design and redesign operations and processes. They are used to give an insight into what is currently happening at different levels of operations within an organization as well as within the supply chain. They are also used to provide decision-makers with a tool to tackle problem areas, the bottleneck in operations and the opportunity to seek innovation. Within the supply chain, performance measures have been developed with the scope to align organizational goals to achieve overall supply chain efficiency, to lead to a high level of customer satisfaction that meets the investment's objectives and respond effectively to a changing environment.

Researchers have discussed extensively the goals that need to be achieved to develop performance measures. Some of these measures are designed with a very specific, narrow scope that only meets the need of a particular operation, process, product or service and uses internal knowledge to develop these measures and readily available data, such as small data. These types of measures may be incorporated into the company's performance measurement system, or they may be a specific project's dedicated measures that have limited use in other situations.

There are also performance measures newly developed that supplement an answer that has not yet been previously provided regarding particular tasks or activities. These types of measures tend to be an add-on to current measures present within the company's performance measurement system. There are also measures developed with the role of benchmarking within or outside an organization with the best in class.

Setting up performance measures is also about understanding what can be measured and what would be the correct 'formula' that would give a clear understanding of what is being represented. The formula referred to in this context is in fact the mathematical formula behind setting up a performance measure. For example, the cost to manufacture a product is represented as a sum of individual operational costs that add up to give an understanding of the final formula used to indicate the final value of the manufacturing cost of a product. One other angle that needs to clearly be portrayed is the visibility of the formula used. For example, if the formula for a set of performance measures is to be noted as: $m1 = c1+c2+c3$ and $m2 = c1*c2*c3$, where $m = measure$, $c = component$, there are clear dependences in terms of the equations constructed. Modifying the equation components will provide us with variations in the proposed measure. If c1 changes its value to 50*c1, m2 will be 50 times its previous value, where m1 will have a slight increase.

Having visibility of the formula used and of variations in data for the components used to form the performance measures will allow for a better understanding of the type of changes to be considered. The expectation is that if we understand the performance measure configuration we can clearly identify changes within the individual components that form the measure. Researchers have invested effort in presenting work that deals with the formulation of measures. For example, Beamon (1999) indicates flexibility type measures and provides formulas to be used for volume flexibility, delivery flexibility measures and mix flexibility reflecting process and job flexibility and new product flexibility, examples of which are presented later in this chapter. Also, Beamon and Chen (2001) provide formulas to be used for resource type measures with specific formulas for average periodic inventory level, average transportation cost; and output type measures with formulas for stock-out fraction and backorder fraction. A number of researchers have used regression and multiple regression methodology to develop formulas for different sets of performance measures (Dubey and Ali, 2015; El-Sakty et al, 2014).

On the one hand, we are dealing with mathematical formulas to be used and developed that clearly reflect a performance measure, but on the other hand, we require accurate data for each individual component that forms the performance measure to obtain a reliable answer. We have noted in the previous chapter the importance of working with accurate, reliable data; now we need to be in a position to work with correct formulas that brings the exact representation of the performance measure being discussed.

A large number of performance measures have been designed over the years, and they will continue to be further developed. Measures were originally designed to provide financial answers such as profit, production cost, return on investment, profit margin, cash flow, etc, but it has been recognized that reporting only on financial performance will not be sufficient to improve the overall efficiency and effectiveness of an organization and the supply chain they are part of. Elgazzar et al (2019) have indicated that in order to provide a balanced assessment of an organization's performance, measures from all categories need to be approached.

Performance measures set internally within an organization do not necessarily reflect the supply chain measures. Lambert and Pohlen (2001) have noted that there is a lack of measures that extend across the supply chain. The tendency noted in developing performance measures was to have these grouped in different categories. Typical categories in which performance measures could be grouped are cost-related measures, time-related measures, reliability measures, responsiveness measures, flexibility measures and measures linked to assets.

Understanding the scope of the measure, the formula to be used in developing the performance measure and the data available to populate the measure to obtain an answer is just the starting point in developing a performance measurement system. Table 4.1 aims to put forward a typical set of questions that can be asked in relation to understanding and setting up performance measures and a performance measurement system within a supply chain setting.

Table 4.1 Sample of supply chain performance measures questions

Questions	Relevant to
How do we know whether something has improved if there isn't a measurable way of showing how or what has improved? (Plenert, 2014, p 42) What are the current performance measures used within the organization? What formula has been used to calculate the individual measures? What are the key variables used to construct the performance measure? Are there any new operational or strategic changes that may invalidate the formula currently being used? Are the results from a current performance measure clear to allow reflection? Which team/individuals are responsible for reviewing current performance measures? How often are current performance measures being evaluated within a department and within an organization? Are measures set on more than cost/finance related characteristics? Which categories of measures have not yet been identified? Is there a performance measurement map that clearly links strategic to operational and financial measures?	Internal to an organization: all working with data, managers responsible for implementing change and innovation; finance managers
What are the performance measures that have been set up to make the link between the suppliers and customers within an organization? What knowledge has been generated as a result of evaluating current performance measure? What validation and verifications tests are carried out to ensure that current measures still reflect changes within an organization? What performance measures require the use of external data? How much does it cost to incorporated external data?	Organizations: strategic managers, operational managers, process managers, data warehouse managers, IT managers; managers responsible for each department within an organization (for example buyers, manufactures, manager responsible for return of goods, delivery manager, service manager, finance manager, HR manager, IT manager, CRM manager)

(continued)

Table 4.1 (Continued)

Questions	Relevant to
	Researchers: in the area of operational management, performance measurement, logistics and supply chain management
What are the steps required to develop a performance measurement system within an organization? What performance measurement system has been developed or is being used within an organization? Is this a supply chain performance measurement system? What measures are being used/developed to reflect the extended supply chain characteristics? Are measures being developed/used related to sustainability? Is the current performance measurement system within an organization capable of capturing sustainability and resilience related measures? Are existing performance measures within an organization capable of capturing fluctuations in behaviour as a consequence of unpredictable disturbances?	Internal to an organization: managers responsible for developing, implementing and operating the organization's performance measurement system, all involved in operating and reporting on performance, IT managers Researchers: in the area of operational management, performance measurement, logistics and supply chain management
Is the current performance measurement system within an organization capable of capturing data-related measures? What are the data performance measures that refer to the aspects of data cost, data reliability, data quality and data accessibility? What prescriptive analytics are being put in place to capture external data that can be incorporated into the current performance measurement system? What advanced analytics, such as optimization simulation, are put in place to confirm optimized performances?	Organizations: managers responsible for developing, implementing and operating the organization's performance measurement system, IT managers Researchers: in the area of operational management, performance measurement, logistics and supply chain management, computer science

The development of supply chain performance measurement systems: Current applications

Balanced scorecard

A large number of performance measurement frameworks and systems have been developed over the years, with some of them presented here.

In 1992 we have the framework developed by Kaplan and Norton as the balanced scorecard (BSC). This framework was intended to create a balanced approach by introducing not only financial measures but also non-financial measures. This framework aims to bring the link between four key aspects of the performances:

- *financial* performance measures;
- performance measures based on *customers'* perspective;
- performance measures from an *internal business* perspective; and
- performance measures from an *innovation and learning* perspective.

It is assumed that, when working with these four categories, an organization can set individual *goals* for each one of these four categories and identify individual *measures* that link to each category. In 1996, Kaplan and Norton further develop the model and considered that based on the company's vision and strategy, these four categories could be set to capture: objectives, measures, targets and initiatives.

With this framework we see the move from financial performances alone, to performances that make the link to the external environment through the customer perspective and also the developmental perspective linked to innovation and learning. The authors of this framework argue that by only considering these four perspectives of measures, it would limit the number of measures used and would give a focus to the organization's strategies. This appears a valid argument and many agree with it (Raval et al, 2019), but with the use of advanced technology and business analytics techniques it can also be argued that an evaluation of more categories of measures that capture the extended supply chain perspective as well as elements of sustainability and resilience would give a more comprehensive view of the situation. Researchers and practitioners have used this framework successfully over a number of years. There are a number of supportive arguments when this framework is applied within an organization, including that it allows a quick adoption of strategic decisions, it is a flexible framework that can be applied to a number of organizations and industries, it allows us to understand customers' views and it promotes a culture of innovation and learning. Bhagwat and Sharma (2007) have successfully develop measures using the BSC approach and apply this to small and medium-sized companies in India. Callado and Jack (2015) consider the use of the balanced scorecard model and attempt to evaluate whether performance measures within this model can be used across the supply chain, using an analysis taken in a Brazilian agri-food supply chain. Bigliardi and Bottani (2010) have developed a tailored BSC model for the food supply chain, and Eskafi et al (2015) have developed a new performance measurement model for the food industry using the BSC in combination with path analysis, evolutionary game theory and cooperative game theory for strategic planning. Supply chain performance measures have been researched by Frederico et al (2020) in the context of Industry 4.0 and proposed a balanced scorecard based on supply chain dimensions identified in the literature.

Some examples of proposed performance measures captured by different authors in these four categories are listed in Tables 4.2–4.5. It should be noted that as the authors have defined the measures following a specific

Table 4.2 Examples of financial measures for the BSC model

Financial measures	Authors (year)
Customer query time, net profit vs productivity ratio, cost per operation, rate of return on investment, buyer–supplier partnership level, delivery performance, delivery reliability, cost per operation hour, variations against budget, information carrying cost, supplier cost savings initiative, supplier rejection rate	Sharma and Bhagwat (2007)
Sales, revenue, cost of human resources, cost of raw materials, goods and external services, other costs, investments, amount of debt	Franceschini et al (2014)
Profitability, liquidity, revenues by product, revenue per employee, contribution margin, level of indebtedness, return on investment, unit cost, minimizing costs, maximizing profits, inventory, overall earnings and operation costs	Callado and Jack (2015)
Increase net profit margin, increase return of equity, meet balance growth goals, deliver expected budget results	Acuña-Carvajal et al (2019)
Shareholder value, level of cost reduction, profitability, earned value added (EVA), earnings before interests, taxes, depreciation and amortization (EBITDA)	Frederico et al (2020)

Table 4.3 Examples of customer measures for the BSC model

Customer measures	Authors (year)
Customer query time, level of customer perceived value of product, range of products and services, order lead time, flexibility of service systems to meet particular customer needs, buyer–supplier partnership level, delivery lead time, delivery performance, effectiveness of delivery invoice methods, delivery reliability, responsiveness to urgent deliveries, effectiveness of distribution planning schedule, information carrying cost, quality of delivery documentation, quality of delivered products, quality of delivered goods, achievement of defect-free deliveries	Sharma and Bhagwat (2007)
Communication, after-sale service, perception of final product/service, organization's image	Franceschini et al (2014)
Customer satisfaction, customer loyalty, new customers, market share, brand value, profitability per customer, revenue per customer, satisfaction of business partners, delivery time, responsiveness to clients, growth in market share, maximizing sales	Callado and Jack (2015)
To restructure the relationship with clients towards long-term commercial links, to systematically deepen understanding of the needs of clients to offer a portfolio of solutions in synch with their business needs, to offer a quality service that generates a high level of satisfaction for clients, to reduce the percentage of deals rejected by clients during negotiation, to increase service level, to grow market share	Acuña-Carvajal et al (2019)
Level of market share, value-added perception, level of customer interaction with processes, level of customer satisfaction	Frederico et al (2020)

category, the measures are in many cases different, which indicates the flexibility on developing these measures. At the same time, if the evaluation is across the supply chain when more than one organization is part of the chain, the communication of the results from these measures may be misleading.

- **Financial measures:** These are set to consider the financial requirements within an organization. They are essential in understanding the organization's financial position on the market. The cost of individual operations is set within this group of performances, where their viability can be identified at this stage. Organizations put particular emphasis on these measures for financial reporting and evaluation (Table 4.2).

Table 4.4 Examples of internal business perspective measures for the BSC model

Internal business perspective measures	Authors (year)
Total supply chain cycle time, total cash flow time, flexibility of service systems to meet particular customer needs, supplier lead time against industry norms, level of supplier's defect-free deliveries, accuracy of forecasting techniques, product development cycle time, purchase order cycle time, planned process cycle time, effectiveness of master production schedule, capacity utilization, total inventory cost, supplier rejection rates, efficiency of purchase order cycle time, frequency of delivery	Sharma and Bhagwat (2007)
Quantitative production level, cycle time, qualitative production level (final products), qualitative production level (incoming products), delivery, stock level, capacity utilization, expansion, satisfaction of human resources, productivity of human resources, security of human resources, environmental impact	Franceschini et al (2014)
Following new products, new processes, productivity per business unit, product turnover, after sales, operational cycle, suppliers, waste, flexibility, response time to customers, delay in delivery, responsiveness of suppliers, storage time, information/integration of materials	Callado and Jack (2015)
To optimize resource management and information processing in the development of internal processes, to make continuous follow-up, analysis and adjustments of loss agreements, to improve effectiveness in the treatment and redirection of external and internal service requests, to structure the process for issuing guarantees and to design a value offer to cater to market needs that arise in the post-conflict era	Acuña-Carvajal et al (2019)
Processes efficiency, response time, level of flexibility, level and extension of transparency, level of collaboration, level of waste reduction, level and extension of processes integration	Frederico et al (2020)

- **Customer measures:** These are set to ensure that any links with the customers are consistent across the organization and they are maintained. Customers are key to any organizations, therefore understanding how they view performances is also paramount (Chytas et al, 2011) (see Table 4.3).
- **Internal business perspective measures:** These are set to evaluate the internal operations, processes and activities of a business and make the

Table 4.5 Examples of innovation and learning measures for the BSC model

Innovation and learning measures	Authors (year)
Supplier assistance in solving technical problems, supplier cost savings initiatives, order entry methods, accuracy of forecasting techniques, service flexibility to meet particular customer needs, range of products and services, level of customer perceived value of products	Sharma and Bhagwat (2007)
Product variety, research and development on products, research and development on process, competitiveness, response time, conformity to customer requirements, rationality in setting and development of projects, education, training and qualification of human resources, self-learning	Franceschini et al (2014)
Investment in training, technology investment, investment in information systems, employee motivation, employee capability, managerial efficiency, employee satisfaction, innovative management, number of complaints, risk management	Callado and Jack (2015)
To narrow the gaps in skill and knowledge at the different levels of the hierarchical structure of the business unit, to redesign functions to decentralize the decision-making process and the interaction with clients, to develop robust and updated systems to handle commercial information, encourage the application and empowerment of the 'business plan' methodology, to promote knowledge and keep a closer relationship with clients, to develop a systematic structure to investigate the environment and predict new business opportunities	Acuña-Carvajal et al (2019)
Adequacy and extension of technologies, adequacy of infrastructure to the new technologies, level of horizontal integration (information and technologies), level of vertical integration (information and technologies), level of people competences, adequacy in meeting compliance and legal requirements, level of leadership engagement, coordination effectiveness	Frederico et al (2020)

link with different functions internally. Creating performances specific to the internal functions will allow managers to identify activities and processes that are not performing as expected. At the same time, internal processes in which the organization is performing very well or over-performing can also be identified. A sample of measures collected from different studies is captured in Table 4.4.

- **Innovation and learning measures:** These explore the impact that learning, training and aspects of innovations will have on the organization's performance (Table 4.5).

As previously mentioned, the way in which these measures are formulated and the way in which they are defined are essential in terms of capturing the intended goal and using these in supporting managers to make decisions on operations and propose change. The measures that are formulated to reach an intended goal, such as 'to increase', 'to decentralize', 'to promote', 'to improve efficiency', may need a set target or a comparable measure to be able to benchmark the results obtained, or an additional measure may be required to understand whether the set measures are really capturing the intended goal.

The BSC framework has a considerable reputation and has been applied in many cases in research and practice and in many industries. However, over the years a number of limitations have been noted regarding this framework. For example in Chytas et al (2011) the following limitations are noted: performance measures in a balanced score card are *placed in a one-way cause-and-effect approach* that could be problematic; the *trade-off between the set measures in the four categories is not taken into consideration*; the *measures are equally weighted* (Acuña-Carvajal et al, 2019), even though in a real situation some measures may carry more weight than others; the *dynamic element is not being captured* when constructing the balanced scorecard. Limitations are also noted in Acuña-Carvajal et al (2019) that refer to the *method of selecting the most important measures* that form part of the model and clearly represent the organization is not clearly determined.

Other aspects that can be noted are: the BSC does not consider an explicit category that captures other extended functions of the supply chain, such as the supplier's perspective. We can see from previous examples (see Tables 4.4 and 4.5) that reference to the supplier's perspective measures are embedded in the internal organization measures and/or aspects of innovation and learning by some authors. However, others do not address this specifically. The BSC does not incorporate a dedicated category related to the reverse aspect in the supply chain, such as the rework, remanufacture, return and reuse aspects. Some organizations may not consider this function at all, where others may refer to the reverse aspect within the customers or innovations sections. It can also be noted that the BSC does not capture elements of environmental and social sustainability, where again they may be embedded within the existing categories if an organization is concerned with these. Limitations of the BSC have also been summarized and captured in the work from Paranjape et al (2006) who also refer to the exclusion of people, suppliers, regulators and competitors. In this study, the authors also mentioned that the environmental and social aspects are not directly captured, and the model is of a static nature.

Supply chain operations reference model

One other well-discussed framework in the literature and implemented within a large number of organizations is the supply chain operations reference model (SCOR), developed in 1996 by the Supply Chain Council (ASCM, 2019) and well cited in the literature by a number of researchers (Elgazzar et al, 2019, 2016, 2012; Ntabe et al, 2015; Ponis et al, 2015; Kocaoğlu et al, 2013). This model gives a standardized way to view the supply chain from the organization's overall plan to individual plans within each department. The other relevant sections incorporated in this model are represented in categories such as: plan, source, make, deliver, return and enable. For each of these categories, performance measures (250 SCOR metrics, ASCM, 2019) are defined and organized into a hierarchical structure with different levels. Measures developed at level 1 consider organizational measures in categories such as reliability, responsiveness, agility, cost and asset management efficiency. Measures at this level are standardized, and when a number of companies within the same supply chain are using these measures, it is evident that they are referring to the same measures and use the same formula. These categories of measures have been explored extensively by many researchers and implemented in many cases. A number of these measures have been listed in Kocaoğlu et al (2013) as level 1 measures that consider a description and a formula for a particular measure.

For example, under the responsiveness category, a measure is: *order fulfillment lead time*, measured in percentages, with a formula give as: *(total orders delivered on time – orders with faulty documentation – orders with shipping damage) / total number of orders received.*

One other example is presented in the assets category, for example a measure is: *cash-to-cash cycle time*, with the measurement unit being time (in days) and the formula used to calculate this measure being: *(inventory days of supply + days of sales outstanding – days of payable outstanding).*

Like these two, measures in each category are well defined and the equation indicating each of the measures is clearly specified, therefore every organization looking to adopt this model will be implementing measures using the same formula for calculating a specific measure in a category. At the same time, when annual reports are being analysed, organization from the same chain that are using the same performance measurement system will report on the exact measures using the same definition for the measure. It is also relevant to indicate that measures in this model are built in a hierarchical form. Measures at level 1 are constructed from measures from previous levels' calculations, such as level 2 and level 3. The calculations of the previous levels come from a narrower number of sub-processes. Level 2 considers process level matrices and level 3 delves into developing diagnostic metrics. Piotrowicz and Cuthbertson (2015) have concluded from their research that both BSC and SCOR models have been considered common approaches among their responses. The popularity of these two models has been discussed in a number of studies that refer to supply chain performance measurement systems (Elgazzar et al, 2019, 2016, 2012). The SCOR model has been applied in many industries; however, some authors comment that some industry specific adaptations may need to be considered. Georgise et al (2017) report that the SCOR model has been applied in the manufacturing industry, construction, service industry, military, geographical information system, information technology, logistics operations, environmental and risk management and collaborative supply chain networks.

Global supply chain forum

One other classical framework is the global supply chain forum (GSCF) (Lambert et al, 1998) with its eight key business processes. These are: customer relationship management (CRM), customer service management (CSM), demand management, order fulfilment, manufacturing flow management, procurement, product development and commercialization and returns (Cooper et al, 1997). The identified functions related to the supply chain have been considered as: *product flow* from tier 2 suppliers, to tier 1 suppliers to manufacturers to customers and end-customers/consumers. The *information flow* function has been indicated together with the purchasing, production, research and development (R&D), finance, marketing and sales and logistics functions.

To understand how CRM affects the economic value added (EVA), Lambert and Pohlen (2001) have put forward measures such as: retain and strengthen relationship with profitable customers, increase sales value, sell higher margin products, improve 'share of customer', improve mix (align services and cost to serve) all within one group referring to sales type measures, and improve plant productivity as cost of goods sold type measures. Within one other group, total expenses, they place measures such as: targeted marketing, reduce services provided to less profitable customers, improve trade spending, eliminate or reduce services provided to low-profit customers, optimize physical network/facilities, leverage new and/or alternative distribution channels, reduce customer service and order management costs, reduce general overhead/management/administrative costs, and reduce human resources costs/improve effectiveness. Under the inventory type category, Lambert and Pohlen identify: improve demand planning, reduce safety stock and make to order, mass customization of inventory. Under fixed assets they have indicated the following three measures: improve asset utilization and rationalization, improve product development and asset investment, and improve investment planning and development; and under other current assets the following is listed: reduce accounts receivable through faster payment. Similarly, the authors have questioned how supplier relationship management affects the economic value added and list similar sets of measures under the same types of categories. Note that reduce finished goods inventory has been listed under the SRM inventory category, where we would also expect this type of measure to be under the CRM.

Researchers have investigated and listed the characteristics of other frameworks that link to performance measures, such as activity-based costing, efficient customer response and economic value added (Estampe et al, 2013; Piotrowicz and Cuthbertson, 2015; Elgazzar et al, 2019).

Having the examples from these well-established frameworks, an organization may decide to focus their attention on developing a tailored supply chain performance measurement model that meets their particular need; or may decide to adopt one of these frameworks, or to adopt a combination of two of the models.

Analytical tools used in connection with performance measures

A number of analytical tools have been considered to work with supply chain performance measurement (SCPM) models, some of which are summarized in Elgazzar et al (2019).

Simulation methodology is used, where a range of performance measures can be designed and implemented in this application. System dynamic modelling can be used with various level of uncertainties incorporated. Jaipuria and Mahapatra (2015) consider uncertainties in demand, supplier lead time, processing time, supplier acquisition rate and others where performance measures such as work in progress inventory, backlog and raw materials shortage have been developed. The software package used in their study is STELLA 5.0. A simulation model supported by mathematical equations is developed in Chan et al (2014) that measures the level of innovation and marketing strategies of a firm. Simulation has also been used for a supply chain case study and quality, lead-time and cost related measures have been developed and analysed in this case (Persson and Olhager, 2002). The software package used to carry out this simulation is Taylor II for discrete event simulations.

Fuzzy logic (Chan and Qi, 2003; Agami et al, 2014); fuzzy-analytical hierarchy processes (fuzzy-AHP) (Elgazzar et al, 2012; El-Baz, 2011); fuzzy Delphi method (Tseng and Liao, 2015); fuzzy decision making trial and evaluation laboratory (DAMATEL) for a BSC with linear programming (Acuña-Carvajal et al, 2019); fuzzy cognitive maps for a proactive BSC (Chytas et al, 2011) are techniques used in relation to performance measures in the supply chain.

Analytical hierarchy process (AHP) has been used by a number of authors and many have used this in combination with another approach, such as fuzzy-AHP (Elgazzar et al, 2012) or AHP – TOPSIS–SCOR (Kocaoğlu et al, 2013), where TOPSIS stands for technique for order preference by similarity to ideal solution. AHP is a technique that is used to help decision-makers to decompose a complex problem into sub-problems all presented in a hierarchical structure. Their application in the field of performance measures research is very popular as measures at different levels can be represented in a hierarchical form, where the objective from one level can feed into the calculations at higher level. AHP has many applications in the field of performance measures, and it has also been applied in areas such as investment appraisal, HR, vendor selection and others (Kocaoğlu et al, 2013).

Optimization approach has been used in combination with other models and approaches such as system thinking, strategic planning, BSC, SCOR and theory of constraints thinking process (TOCTP) to developed a comprehensive framework of performance measures (Agami et al, 2012). A mixed integer linear programming model has been developed by Sabri and Beamon (2000) and they consider as their objective the following measure: minimization of fixed and variable cost for components such as: raw material purchase price and transportation from vendors to plants; cost of handling products at the DC and transportation of products from plants to DCs; cost associated with plant operations; and the transportation cost of products from DCs to customers. Many models using linear programming are developed with the purpose to operate with cost-based measures, where other associated performance measures could also be developed within these models. A number of examples using optimization of performance using linear programming are detailed in Chapter 7.

Statistical process control (SPC) is a visual tool that uses an upper and lower control limit as a normal process variation with a central line as an average and this is used to indicate the variation that exists in the system under study. Morgan and Dewhurst (2007) have compared the use of SPC control charts with descriptive statistical analysis to evaluate the supplier–buyer relationship and observe the effect of a supplier's performance. The SPC provides performance monitoring and the opportunity for problem-solving discussions.

Data envelopment analysis (DEA) models have been used by many researchers in performance measures research to measure efficiency of processes or systems (Wong and Wong, 2007, Ozbek et al, 2012; Agrell and Hatami-Marbini, 2013; Nwanosike et al, 2016). DEA is a mathematical technique that uses as a base the principles of linear programming and is considered relevant by many authors as it has the ability to use a number of input measures to generate outputs based on decision-making units (DMUs) of measures established. DEA techniques have been used by researchers in combination with another model to enhance its effect. Wang et al (2016) combine DEA with the BSC to evaluate the impact of operational efficiency on overall performance when using financial and non-financial measures. Hong and Jeong (2019) have used the DEA in combination with a multi-objective model to evaluate the efficiency of facility allocation decisions.

Statistical models have been used by a number of researchers in the area of performance measurement. For example, Dubey and Ali (2015) and Beamon and Chen (2001) use multiple regression analysis for studies in supply chain performance measures. In many other studies we see the use of a statistical model alone or in combination with other models.

The future of performance measurement systems

Designing performance measurement systems has been extensively covered in the literature. However, practical examples of successful implementations of performance measurement systems are still limited in practice.

Mathematical model used for set measures

An existing set of measures, or an existing measurement system that organizations are working with, may have been developed in the past and may comprises mathematical formulas that do not entirely capture advancements in processes, operations or new business ventures. Once the mathematical formulas have been set for an individual performance measure this does not imply that there is no need to change them. The formulas used in the set measures need to be continuously monitored and updated to clearly reflect changes in business processes and operations.

Capturing big data in the measurement system

We have noted the opportunities and challenges offered by data and the availability of big and small data from previous chapters (see Chapter 3). In many cases, the generation of data and data available outside an organization may not end up in any measures that are captured in the current measurement system an organization is using. This may be because the data generated is unreliable and inconsistent, or it may be that the users are not aware that this data is being generated. Data generated outside an organization and without a clear aim may bring significant benefits and provide a different angle to understand a particular situation, but without a predefined goal it will not be incorporated into a measurement system. Identifying ways to incorporate big data into the measurement system of an organization is one key task for further development in the area of performance measurement systems.

Understanding the link between measures

As part of the same system, a process change will have an effect not only on one particular set measures, but also on others. Understanding the connection between the measures and the effect a change in a process or system has on other individual measures is part of an analysis of the overall organization's performance. This aspect may not be directly obvious if measures are not being constructed yet, or they do not capture a clear representation of the system under study. Statistical analyses have been used to understand the correlation between different measures in the system; however, one other approach could be the use of simulation. Processes within an organization can be simulated, where performance measures can be developed to monitor the performance of individual processes. Using simulations, changes to processes and the way measures are being affected over time could be observed using visualization boards within a simulation.

Operate real-time performance measurement systems

Using appropriate visualization tools, reliable approaches from the point of view of process control and quality evaluation, allows the evaluation of performance measures in real time, and therefore managers will be able to react more quickly to process and system changes. In this case, measures need to be set up to key points in the supply chain for collecting data where manual as well as artificial tools can be embedded to control the process based on the established performance target.

Summary

This chapter looked at different models presented in the literature as performance measurement models, such as the BSC, SCOR and the GSCF, that have been applied in many different examples in supply chains. A number of analytical models are used in connection with performance measures and some of them have been captured in this chapter. Performance measures can only be used in connection with reliable data, and the way in which the data is being collected should be guided by the way performance measures are being defined.

References

Acuña-Carvajal, F, Pinto-Tarazona, L, López-Ospina, H, Barros-Castro, R, Quezada, L and Palacio, K (2019) An integrated method to plan, structure and validate a business strategy using fuzzy DEMATEL and the balanced scorecard, *Expert Systems with Applications*, 122, 351–68

Agami, N, Saleh, M and Rasmy, M (2012), A hybrid dynamic framework for supply chain performance improvement, *IEEE Systems Journal*, 6(3), 469–78

Agami, N, Saleh, M and Rasmy, M (2014) An innovative fuzzy logic based approach for supply chain performance management, *IEEE Systems Journal*, 8(2), 336–42

Agrell, P J and Hatami-Marbini, A (2013) Frontier-based performance analysis models for supply chain management: State of the art and research directions, *Computers and Industrial Engineering*, 66(3), 567–83

ASCM (Association for Supply Chain Management) (2019) Understand the structure of SCOR, www.apics.org/apics-for-business/benchmarking/scormark-process/scor-metrics (archived at https://perma.cc/N7N5-NMNS)

Beamon, B M (1999) Measuring supply chain performance, *International Journal of Operations and Production Management*, 19, 275–92

Beamon, B M and Chen, V C P (2001) Performance analysis of conjoined supply chains, *International Journal of Production Research*, 39, 3195–218

Bhagwat, R and Sharma, M K (2007) Performance measurement of supply chain management: A balanced scorecard approach, *Computers and Industrial Engineering*, 53, 43–62

Bigliardi, B and Bottani, E (2010) Performance measurement in the food supply chain: A balanced scorecard approach, *Facilities*, 28(5–6), 249–60, doi:10.1108/02632771011031493

Callado, A A C and Jack, L (2015) Reflective practice balanced scorecard metrics and specific supply chain roles, *International Journal of Productivity and Performance Management*, 64(2), 288–300, doi:10.1108/IJPPM-05-2014-0071

Chan, F T S and Qi, H J (2003) An innovative performance measurement method for supply chain management, *Supply Chain Management: An international journal*, 8(3), 209–23

Chan, F T S, Nayak, A, Raj, R, Chong, A Y L and Manoj, T (2014) An innovative supply chain performance measurement system incorporating research and development (R&D) and marketing policy, *Computers and Industrial Engineering*, 69, 64–70

Chytas, P, Glykas, M, and Valiris, G (2011) A proactive balanced scorecard, *International Journal of Information Management*, 31(5), 460–468, doi: 10.1016/j.ijinfomgt.2010.12.007

Cooper, M C, Lambert, D M and Pagh, J D (1997) Supply chain management: More than a new name for logistics, *The International Journal of Logistics Management*, 8, 1–14

Dubey, R and Ali, S S (2015) Exploring antecedents of extended supply chain performance measures, *Benchmarking: An international journal*, 22, 752–72

El-Baz, A M (2011) Fuzzy performance measurement of a supply chain in manufacturing companies, *Expert Systems with Applications*, 38(6), 6681–88

El-Sakty, K, Tipi, N, Hubbard, N and Okorie, C (2014) The development of a port performance measurement system using time, revenue and flexibility measures, *International Journal of Business and General Management*, 3(4), 17–36

Elgazzar, S, Tipi, N S, Hubbard, N J and Leach, D Z (2012) Linking supply chain processes' performance to a company's financial strategic objectives, *European Journal of Operational Research*, 223, 276–89

Elgazzar, S H, Tipi, N S and Hubbard, N J (2016) The impact of supply chain strategy on the financial performance: A case study of a manufacturing company, 21st International Symposium on Logistics (ISL 2016), Kaohsiung, Taiwan

Elgazzar, S, Tipi, N and Jones, G (2019) Key characteristics for designing a supply chain performance measurement system, *International Journal of Productivity and Performance*, doi:10.1108/IJPPM-04-2018-0147

Eskafi, S H, Roghanian, E and Jafari-Eskandari, M (2015) Designing a performance measurement system for supply chain using balanced scorecard, path analysis, cooperative game theory and evolutionary game theory: A case study, *International Journal of Industrial Engineering Computations*, 6(2), 157–72

Estampe, D, Lamouri, S, Paris, J and Djelloul, S (2013) A framework for analysing supply chain performance evaluation models, *International Journal of Production Economics*, 142(2), 247–58

Franceschini, F, Galetto, M and Turina, E (2014) Impact of performance indicators on organisations: A proposal for an evaluation model, *Production Planning and Control*, 25(9), 783–99, doi:10.1080/09537287.2012.756128

Frederico, G F, Garza-Reyes, J A, Kumar, A and Kumar, V (2020) Performance measurement for supply chains in the Industry 4.0 era: A balanced scorecard approach, *International Journal of Productivity and Performance Management*, 19, doi:10.1108/ijppm-08-2019-0400

Georgise, F B, Wuest, T and Thoben, K D (2017) SCOR model application in developing countries: Challenges and requirements, *Production Planning and Control*, 28(1), 17–32

Hong, J D and Jeong, K Y (2019) Combining data envelopment analysis and multi-objective model for the efficient facility location–allocation decision, *Journal of Industrial Engineering International*, 15(2), 315–31, doi:10.1007/s40092-018-0294-2

Jaipuria, S and Mahapatra, S S (2015) Performance improvement of manufacturing supply chain using back-up supply strategy, *Benchmarking: An international journal*, 22(3), 446–64

Kaplan, R S and Norton, D P (1992) The balanced scorecard: Measures that drive performance, *Harvard Business Review*, 70, 71–79

Kaplan, R S and Norton, D P (1996) Using the balanced scorecard as a strategic management system, *Harvard Business Review*, 74(1), 75–85

Kocaoğlu, B, Gülsün, B and Tanyaş, M (2013) A SCOR based approach for measuring a benchmarkable supply chain performance, *Journal of Intelligent Manufacturing*, 24(1), 113–32

Lambert, D M and Pohlen, T L (2001) Supply chain metrics, *The International Journal of Logistics Management,* 12(1), 1–19

Lambert, D M, Cooper, M C and Pagh, J D (1998) Supply chain management: Implementation issues and research opportunities, *The International Journal of Logistics Management*, 9(2), 1–19

Morgan, C and Dewhurst, A (2007) Using SPC to measure a national supermarket chain's suppliers' performance, *International Journal of Operations and Production Management*, 27(8), 874–900, doi:10.1108/01443570710763813

Ntabe, E N, LeBel, L, Munson, A D and Santa-Eulalia, L A (2015) A systematic literature review of the supply chain operations reference (SCOR) model application with special attention to environmental issues, *International Journal of Production Economics*, 169, 310, doi:10.1016/j.ijpe.2015.08.008

Nwanosike, F, Tipi, N S and Warnock-Smith, D (2016) Productivity change in Nigerian seaports after reform: A Malmquist productivity index decomposition approach, *Maritime Policy and Management*, doi:10.1080/03088839.2016.1183827

Ozbek, M E, de la Garza, J M and Triantis, K (2012) Efficiency measurement of the maintenance of paved lanes using data envelopment analysis, *Construction Management and Economics*, 30(11), 995–1009, doi:10.1080/01446193.2012.725939

Paranjape, B, Rossiter, M and Pantano, V (2006) Performance measurement systems: Successes, failures and future: A review, *Measuring Business Excellence*, 10(3), 4–14

Persson, F and Olhager, J (2002) Performance simulation of supply chain designs, *International Journal of Production Economics*, 77, 231–45

Piotrowicz, W and Cuthbertson, R (2015) Performance measurement and metrics in supply chains: An exploratory study, *International Journal of Productivity and Performance Management*, 64(8), 1068–91

Plenert, G J (2014) *Supply Chain Optimization Through Segmentation and Analytics*, Volume 48, CRC Press, Boca Raton, FL

Ponis, S T, Gayialis, S P, Tatsiopoulos, I P, Panayiotou, N A, Stamatiou, D R I and Ntalla, A C (2015) An application of AHP in the development process of a supply chain reference model focusing on demand variability, *Operational Research*, 15(3), 337–57, doi:10.1007/s12351-014-0163-8

Ramanathan, U (2014) Performance of supply chain collaboration: A simulation study, *Expert Systems with Applications*, 41(1), 210–20

Raval, S J, Kant, R and Shankar, R (2019) Benchmarking the Lean Six Sigma performance measures: A balanced score card approach, *Benchmarking: An international journal*, 26, 1921–47

Sabri, E H and Beamon, B M (2000) A multi-objective approach to simultaneous strategic and operational planning in supply chain design, *Omega*, 28(5), 581–98

Sharma, M K and Bhagwat, R (2007) An integrated BSC–AHP approach for supply chain management evaluation, *Measuring Business Excellence*, 11(3), 57–68, doi:10.1108/13683040710820755

Teimoury, E, Chambar, I, Gholamian, M R and Fathian, M (2014) Designing an ontology-based multi-agent system for supply chain performance measurement using graph traversal, *International Journal of Computer Integrated Manufacturing*, 27(12), 1160–74, doi:10.1080/0951192x.2013.874584

Tseng, P H and Liao, C H (2015) Supply chain integration, information technology, market orientation and firm performance in container shipping firms, *International Journal of Logistics Management*, 26(1), 82–106, doi:10.1108/IJLM-09-2012-0088

Wang, C H and Chien, Y W (2016) Combining balanced scorecard with data envelopment analysis to conduct performance diagnosis for Taiwanese LED manufacturers, *International Journal of Production Research*, 54(17), 5169–81, doi:10.1080/00207543.2016.1156780

Wong, W P and Wong, K Y (2007) Supply chain performance measurement system using DEA modelling, *Industrial Management and Data Systems*, 107(3), 361–81

PART TWO
Advances in supply chain analytics and modelling

Visualization techniques in the supply chain 05

LEARNING OBJECTIVES

- Understand the power of data visualization.
- Be able to ask and reflect on questions related to visualizing data and processes in the supply chain.
- Understand how data, processes and information can be visualized using different tools.

Introduction

Visualization is a very powerful tool that aims to bring further clarity to the data available and to the analysis carried out on the data, and it can provide a new interpretation of what is in front of us. In other words, visualization tools help us to make invisible aspects within the data more visible. With the use of technology in everyday life and media, we have become used to looking at data through visual interfaces, and this has become an expectation. We expect to observe information changing over time, using a line graph, or see aggregated information on, for example, annual cost per product through a bar chart, and so on. Visual tools and graphs are embedded in a number of publications we see every day, in media communication and various other displays of data and they are also incorporated in a number of educational programmes from an early stage. Therefore, we are becoming more and more familiar with the use of these tools and our expectation that data will be provided in a well-structured format is becoming the norm.

Visualization puts facts into perspective and allows us to take different views on how to start an analysis and/or what to analyse next. Visualization allows us to see continuity in data, missing values, or peaks and valleys in data that we may not have been able to observe otherwise.

There are a number of reasons why we seek more and more visual tools and platforms to represent data, information and processes:

- A large amount of data and information is available to us in many different forms and formats, from data in numerical format, where the classical graphs are used, to information presented in a more structural setting, where particular models are developed and used to support our understanding and enhance the decision-making process.
- The use of visual tools like graph and charts could help identify characteristics in data that are not easy to spot from the raw data. For example, they can highlight errors in input data where values have been missed, or incorrectly entered in the system, or identify a cluster or a linear aspect to the data behaviour.
- The opportunity to share data and analyse data is much easier, and this is also a reason for seeking to work with and employ more visual tools in representing data and information. Many of the software packages we consider and use for collecting and analysing data have easy-to-use functions that allow a non-computer-specialist to use them and transfer evaluations and results into different documents and formats to relate the information. The communication of data is much easier, not only in terms of being able to create a graph in one software package and transfer this easily into another, but also when working with much more aggregated data.
- Within a supply chain, there are a number of points where data is being generated, where visual tools provide the means to compare data at different points and check for variations that will allow decision-makers to implement corrective actions.

This chapter tries to characterize the power of visualization and brings forward a few techniques that have direct applicability in the supply chain setting. The aim in this case is to look at visualizing data for the purpose of analysis, but we also take a critical look at structures in the supply chain and how these could be visualized. This chapter is not, however, looking to describe various developed algorithms that recognize objects and transform data into visual elements for recognitions and simulation, but it is recognized here that these form an important part of logistics and supply chain operations.

A number of software packages can be used to visualize information, data and structural representation relevant in the field of supply chain. A number of data visualization examples used in this chapter are represented using Excel and Minitab. For process-type information, flow diagrams and value stream maps and software packages such as Visio and PowerPoint are also used.

Data visualization

The data visualization concept relates to the use of data in a 'picture' type format for a better understanding, and in many cases to give a different view to the same content. However, the initial question is, how best can the data available to us be represented? This question is present in many meetings, in many situations, and is aiming to relate information and details in the most succinct way possible. In some situations, such as in a meeting, or in a face-to-face class discussion forum, we may not have a software package available to us to start using, and a sketch is the first attempt to represent information or data. In many cases we see that the manual skills in constructing graphs and representing data is overlooked. The tools needed include colour pens, paper, graph paper, ruler, compass and so on. Using these tools requires particular skills and conventions that are in general automated when computer software is used. In many cases, with the use of software packages that display data automatically, the skills required to represent data using manual tools (see Figure 5.1) are no longer or very little in use.

Figure 5.1 Manual representation of data

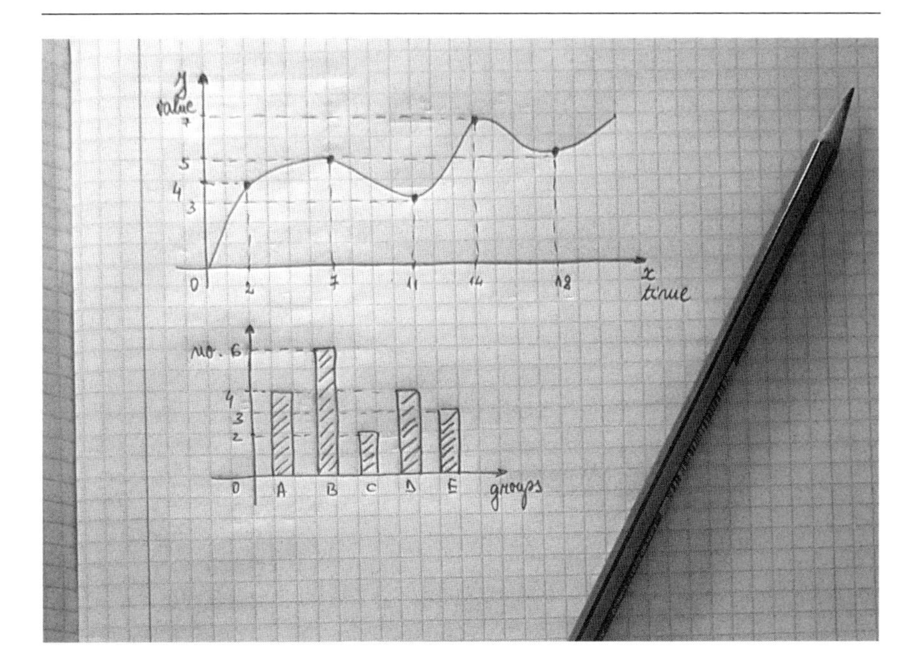

Constructing a diagram or a figure consists of: a title, and in some cases a subtitle; the representation of the x and y axis with a clear indication of what they represent; particular labels that describes the different representation of the data considered; space to represent the data, which is usually in the middle of the diagram. The diagram or figure will also have a name, and if this is from a particular source the source is also indicated. In many cases we see the forecasts for particular products represented on a graph, where the x coordinate represents the 'time', and the y coordinate represents the 'value' of data. More than one forecast can be considered on one graph, where the difference will be indicated with particular 'labels'.

A good understanding of the graphs and their individual components and the process of how to construct and represent them will allow us to understand and describe the key concepts of the data visually. Visualizing the data available in different formats, and using different representations, could give a new angle to the data available, and in many cases aims to be used as a support tool to enhance the decision-making process.

With the advancement in technology and the availability of a large number of software packages that allow for the construction of high-quality, complex graphs and graphical representations, the production of these graphs has become easier and easier. Organizations are storing large sets of data in their internal databases, but data in these formats are only useful if processed and analysed (Sahay, 2017). To visualize data in a picture type format, we often have a range of options available such as: scatter diagrams, line graphs, pie charts, bubble charts and may others.

Data visualization using Excel

If we are working with numerical data, and we have the sales data of a number of products, they can be represented in Excel in the following ways.

Spreadsheet

In a spreadsheet type format, data is listed in rows or columns, as depicted in Figure 5.2. One limitation in this case is that the volume of data that can be visualized in a spreadsheet is limited to the screen size available. However, spreadsheets have the advantage of being able to hold a large volume of data in many rows and within many columns. For large volumes of data, the visual representation of the data in question can be very small in size and it loses value in its interpretation due to the difficulty of observing changes and fluctuations in data represented. However, the representation

Figure 5.2 Spreadsheet representation of sample data

	A	B	C	D
1	**NT Sales Data**			
2	Month	Product NT1	Product NT2	Product NT3
3	January	35	187	320
4	February	28	120	322
5	March	10	98	334
6	April	4	89	354
7	May	28	99	358
8	June	33	105	367
9	July	59	110	361
10	August	87	120	370
11	September	95	132	378
12	October	94	98	381
13	November	120	89	388
14	December	150	110	390

of data in this format provides a lot of details as it highlights the exact value (in Figure 5.2, the sales value per month for each products) and can be used in analysis as a key reference point.

Data in Figure 5.2 is presented in Excel, where this is simply keyed in, or it can be copied from another file format or imported from a different database. The same set of data can be easily transferred in another software package that uses a spreadsheet representation, such as Minitab (Figure 5.11).

Line graph

Data can be represented as a line graph. In Figure 5.3 we see a line graph for each product data detailed in Figure 5.2 as well as a line graph representation for all three products.

To obtain the graphical representation in Excel the following steps can be followed:

- Select the sales data from column B, C and D (from cell B2 to cell D14).
- Go to the tab Insert and select Charts (as shown in Figure 5.4). At this point we note that in Excel there are a number of options that could help us in constructing the graph (Column, Line, Pie, Bar, Area, Scatter, Stock, Surface, Radar and Combo), some of which will be discussed in this chapter. For a line graph a few options are available (such as Line, Stacked Line, 100% Stacked Line, Line with Markers, Stacked Line with Markers, 100% Stacked Line with Markers and 3-D Line). By selecting Line, a line graph as indicated in Figure 5.3 will be displayed, but the lines selected

Figure 5.3 Line graph representation of sample data

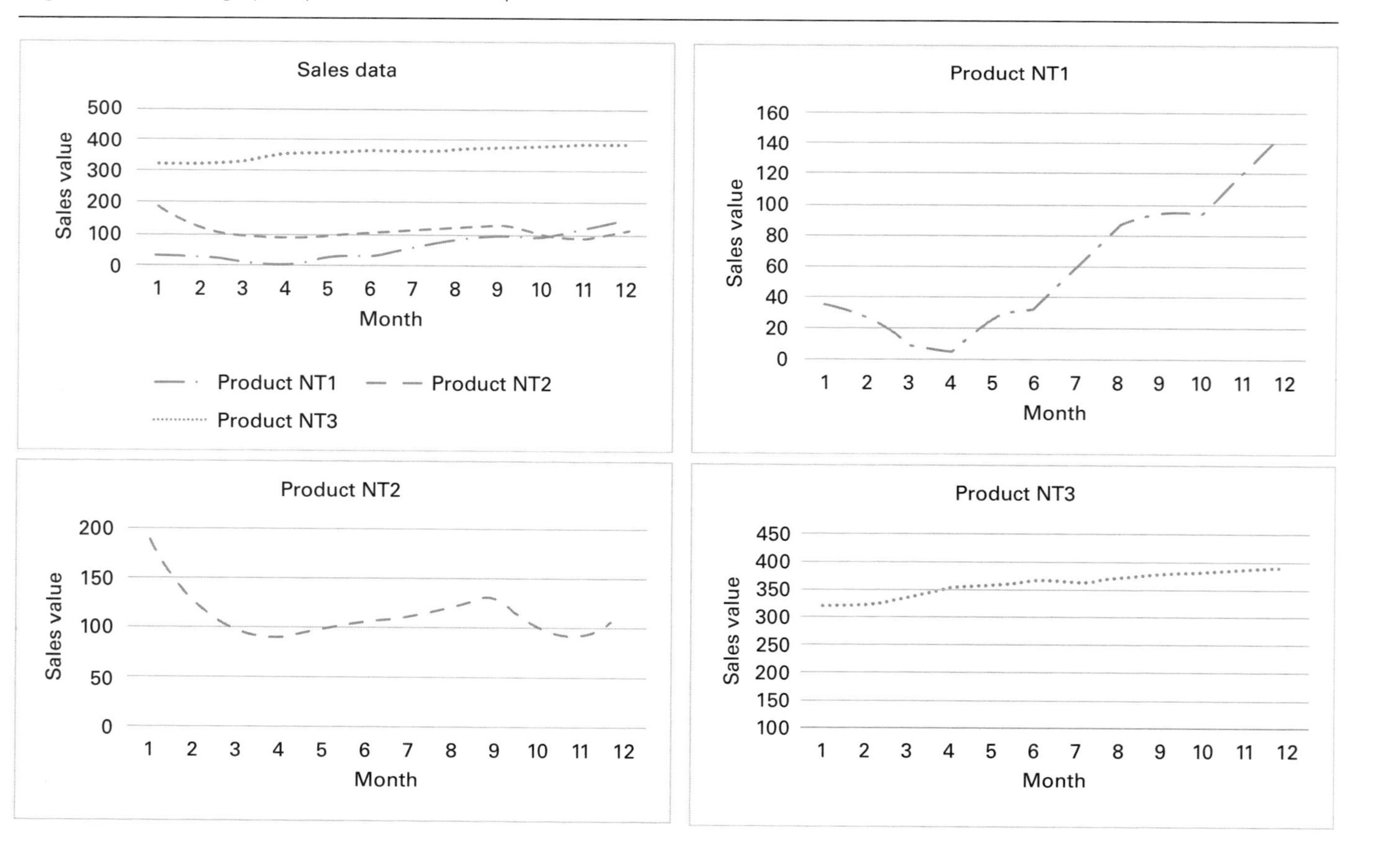

Figure 5.4 Inserting a line chart in Excel

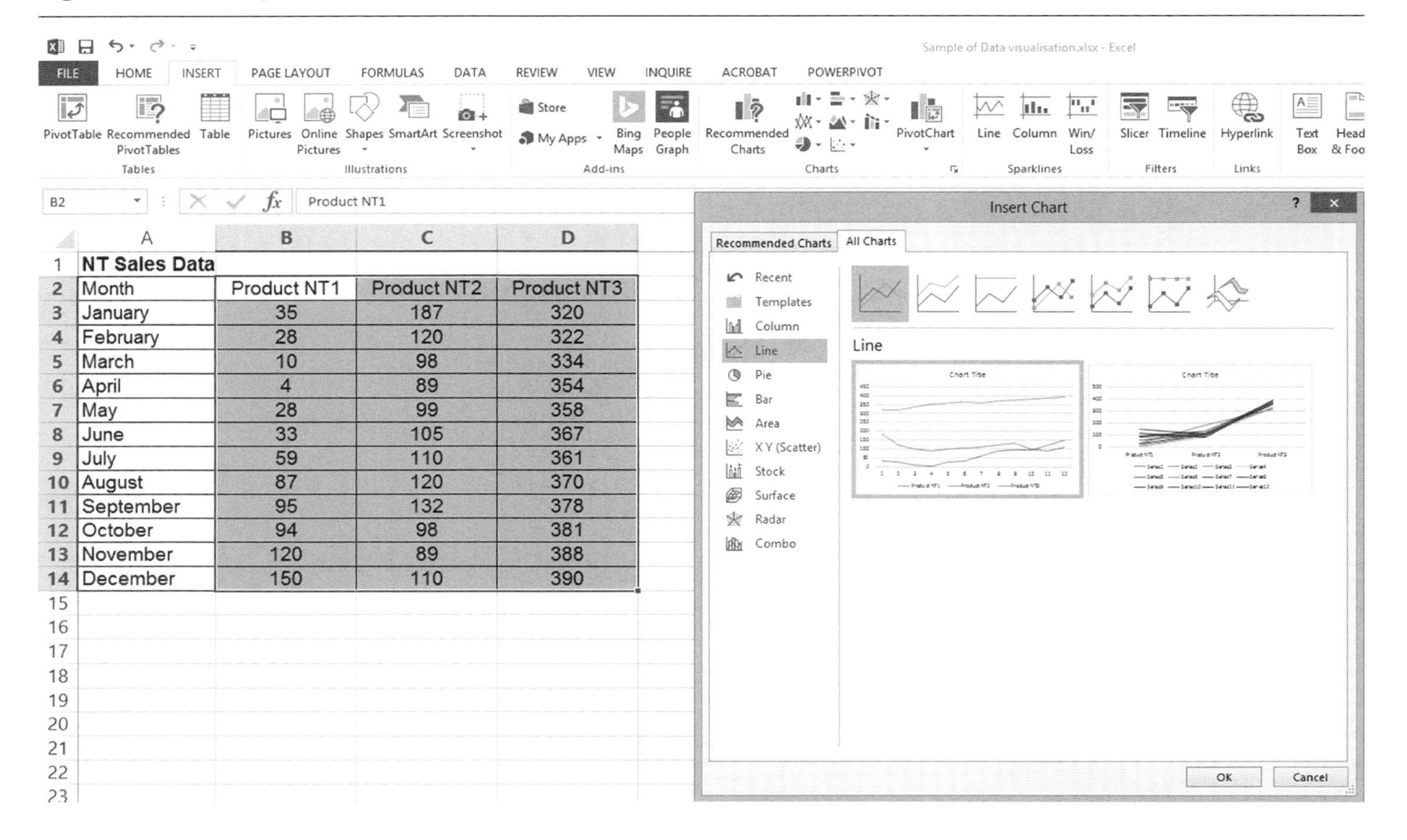

	A	B	C	D
1	NT Sales Data			
2	Month	Product NT1	Product NT2	Product NT3
3	January	35	187	320
4	February	28	120	322
5	March	10	98	334
6	April	4	89	354
7	May	28	99	358
8	June	33	105	367
9	July	59	110	361
10	August	87	120	370
11	September	95	132	378
12	October	94	98	381
13	November	120	89	388
14	December	150	110	390

may have a default selection of an uninterrupted line with a particular colour. However, there is the option to make changes to the representation of the line and format of the graph.

- Select OK, and position the graph in the best place on the screen for conducting the analysis.

Individual graphs can also be represented from the data displayed, by selecting a particular column and following the same steps to obtained the representation of a line graph for one product, as depicted in Figure 5.3.

An advantage in the figure comprising the data value of all three products is that this gives a comparison figure between each product selected for analysis. The details of fluctuations in data are better seen in the figures that focus on individual products. It is evident from the representation of data for product 1, that the sales are decreasing up to month 4 (April) followed by a steady increase to month 12 (December). This graphical representation is particularly relevant when conducting time-series forecasting analysis (see Chapter 6) for different products.

Column chart

A similar representation can be attributed to data from Figure 5.2 in column charts (see Figure 5.5). This representation of data, however, indicates the same values as in Figure 5.3, but it may bring new understandings of the data represented.

To obtain this graphical representation of the data, the same steps can be followed as in the case of a line graph, but the graph selected is now a column chart. There are a number of column charts available for selection, as indicated in Figure 5.5, such as Clustered Column chart, Stack Column, or 100% Stacked Column, where data is represented from the overall percentage and four other selections in 3-D.

Stack bar chart

Data represented as a stack bar chart (see Figure 5.6) brings a new angle to visualizing the data per month. The overall sales value can be noted in this case per month and therefore an analysis of which is the month with the lowest sales value and the highest value can be observed from this representation. In our case the month with the highest sales value is obtained in December, where the lowest is in March.

Figure 5.5 Column charts for a sample of data

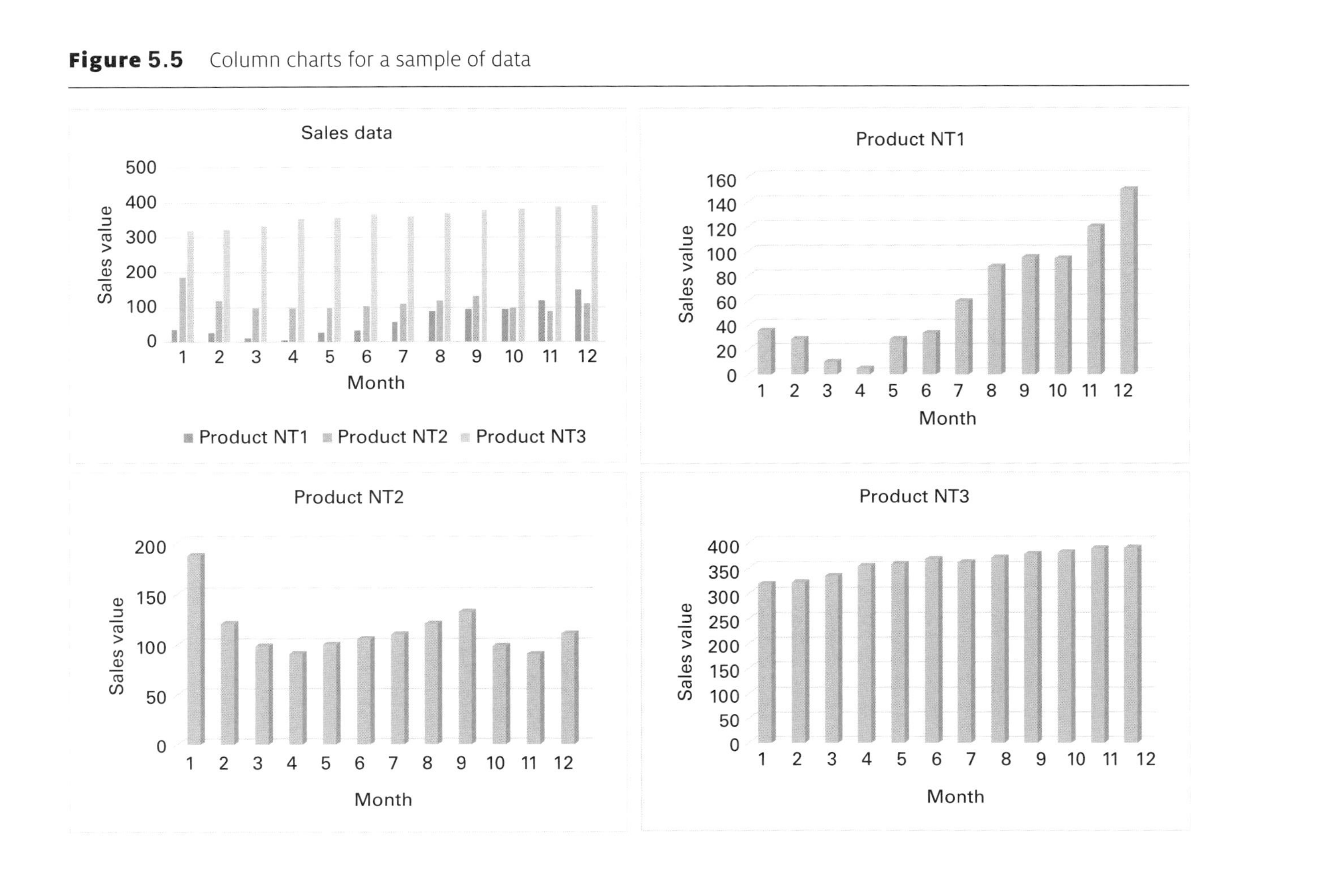

Figure 5.6 Stack bar chart representation of sample data

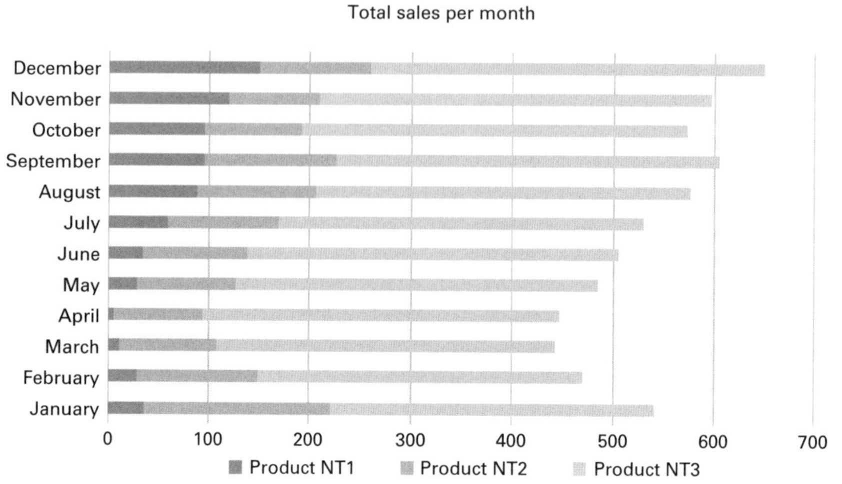

Total sales per month

December
November
October
September
August
July
June
May
April
March
February
January
0 100 200 300 400 500 600 700
Product NT1 Product NT2 Product NT3

Pie chart

A pie chart representation of data is also seen in many analyses. This is relevant when looking to understand our data in relation to the total, where the circle representation (the pie) of the data indicates either the total value or 100 per cent. In Figure 5.7 the total annual sales for each product is represented in the annual sales chart. This shows that the highest sales are formed of the product NT3, where the lowers value sales are obtained from the product NT1. Going further, we also see a sample of data for individual months (such as January, March and December).

Radar chart

To work with a radar chart, data can be collected for four products based on a number of categories, as indicated in Figure 5.8. Data in this example has been collected for four products based on a set of predefined categories. The values within each category range from 1 to 6, where 1 represents the lowest value in each category and 6 the highest value. All products can be represented in one graph. However, as the number of products increases, the representation in one chart may be very difficult to distinguish.

Figure 5.7 Pie chart representation of sample data

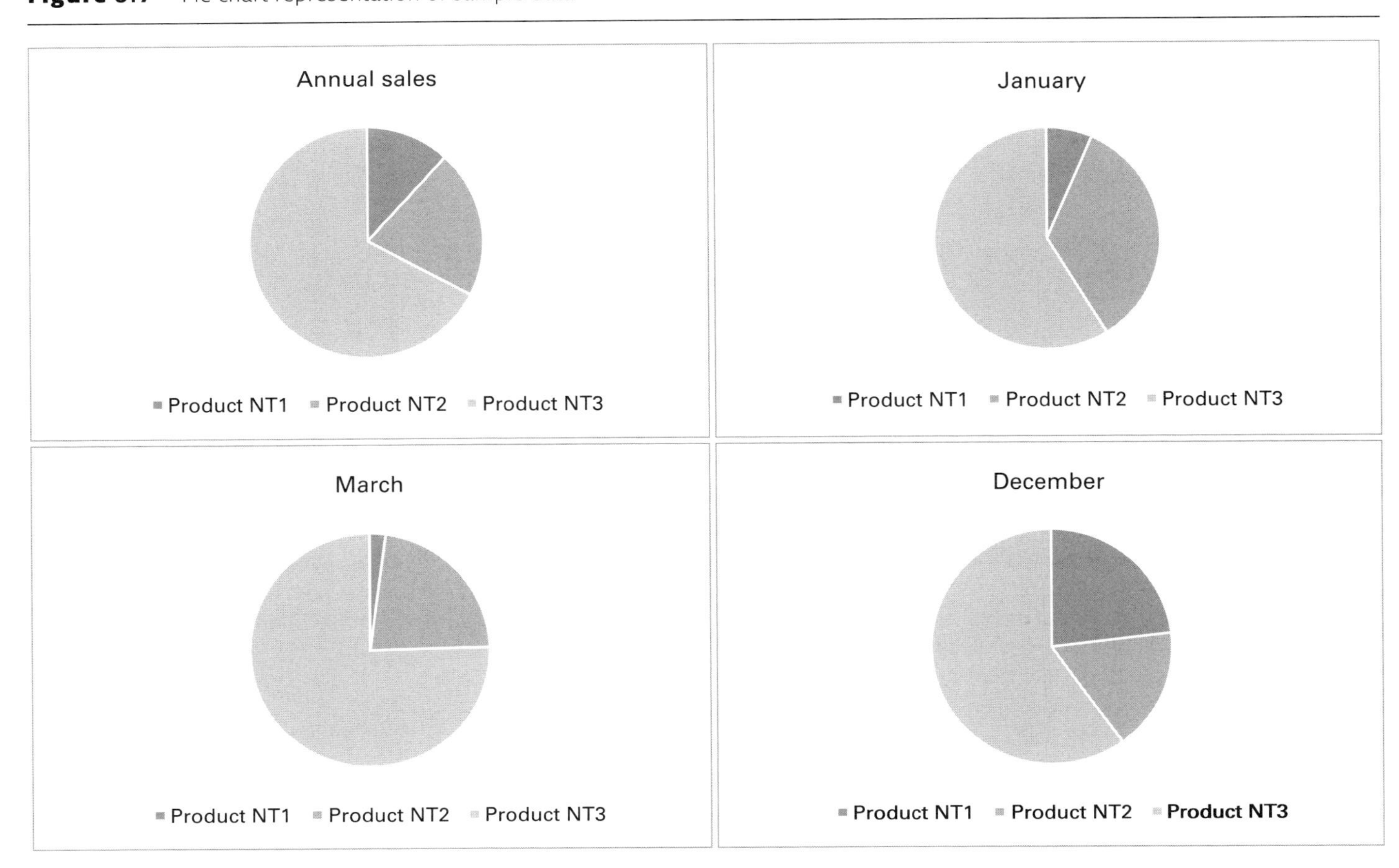

Figure 5.8 Radar chart for a sample of data

	A	B	C	D	E
1	**Categories**	Product NT11	Product NT12	Product NT13	Product NT14
2	Quality	2	4	6	1
3	Usability	6	6	3	4
4	Cost	6	5	4	3
5	Env. Sustainability	2	4	5	6
6	Easy to handle	1	3	6	6

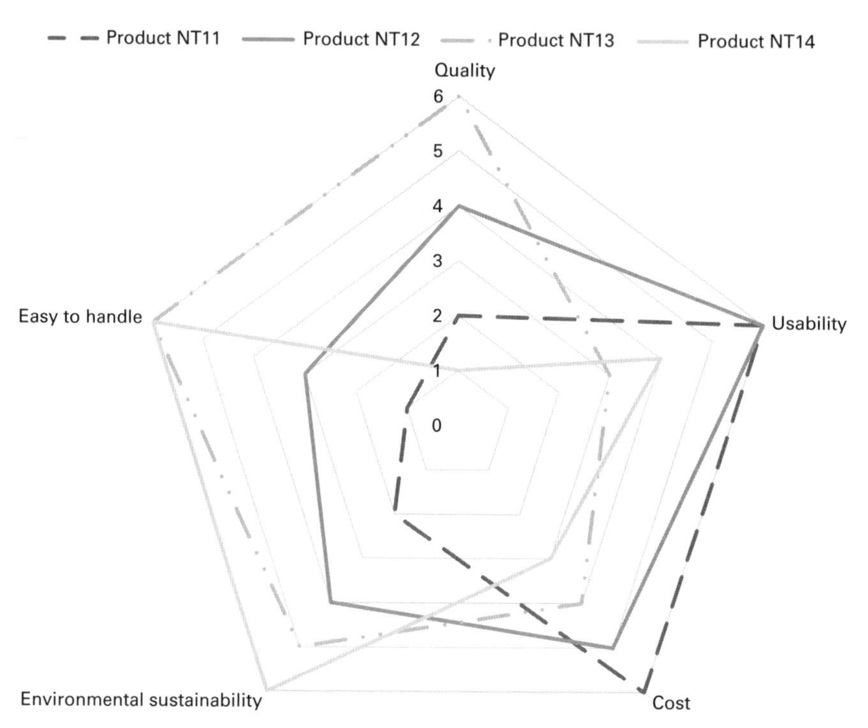

Figure 5.9 Radar chart for one product

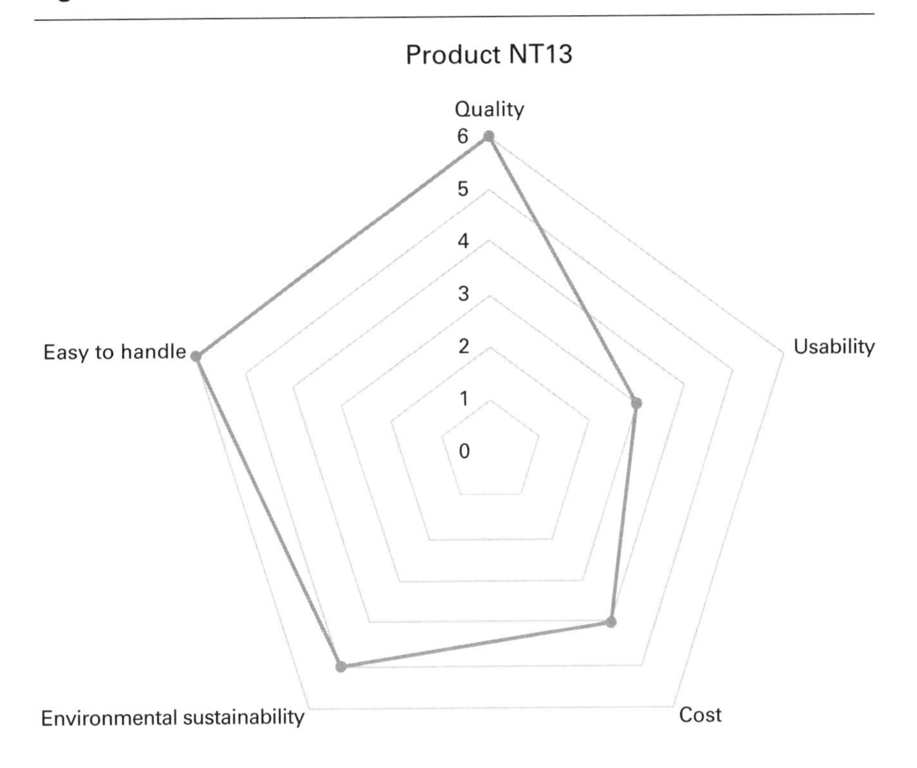

Data characterizing only one product can be represented alone, when an analysis is to be carried out only on one product based on the predefined categories (Figure 5.9).

Gantt chart

Gantt charts indicate the link between activities within a project, the length of each activity, the project finishing time, and whether there are any activities that are happening at the same time. These types of charts are very useful to managers and decision-makers, as they can indicate whether a project and its activities are on track. An example of a Gantt chart developed in Excel is given in Figure 5.10.

Data visualization using Minitab

Minitab is primarily concern with data analysis. However, it does have a number of data visualization functionalities. Below a few examples are

Figure 5.10 Example of a Gantt chart

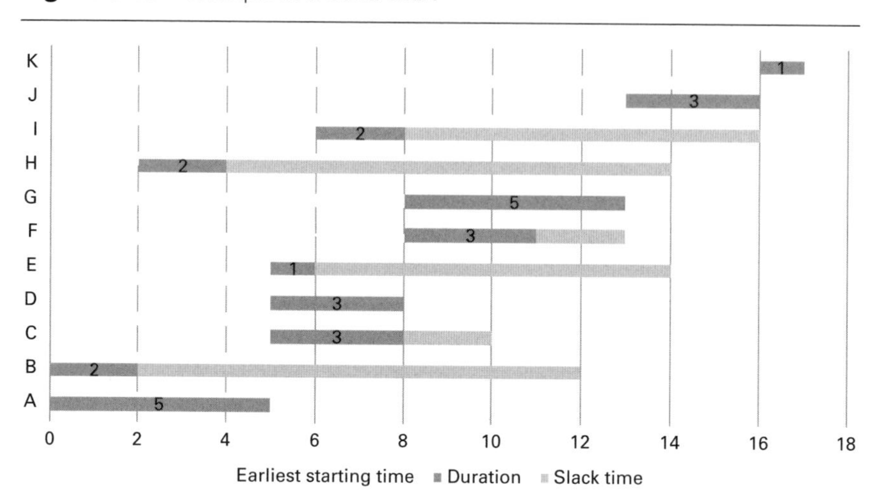

given. Following from the same set of data presented in Excel, we will now look at presenting this using Minitab. The same set of data can be cut and pasted in Minitab in the spreadsheet view (Figure 5.11).

To construct a line graph in Minitab similar to those developed using Excel (Figure 5.3), the following set of instructions can be followed.

Graph – Time Series Plot... and at this point if *Simple* is selected followed by *OK*, the representation of one product over a period of the time available is displayed (Figure 5.12).

As well as displaying each set of data individually using a line graph, or in this case a time series graph, multiple sets of data can also be presented on a graph following: *Graph – Time Series Plot... – Multiple* and select all the products in the *Series*. This will obtain a graph as in Figure 5.13.

An area graph for each product in Minitab can be presented as: *Graph – Area Graph* – and all products can be selected at this point (see Figure 5.14).

Similarly, bar charts can also be created and displayed in Minitab following: *Graph – Bar Charts...* – from here, select *Values From a Table* under the *Bars Represent* (see Figure 5.15). Still considering bar charts, a comparison display can be represented in Minitab as given in Figure 5.16.

Figure 5.11 Minitab data visualization

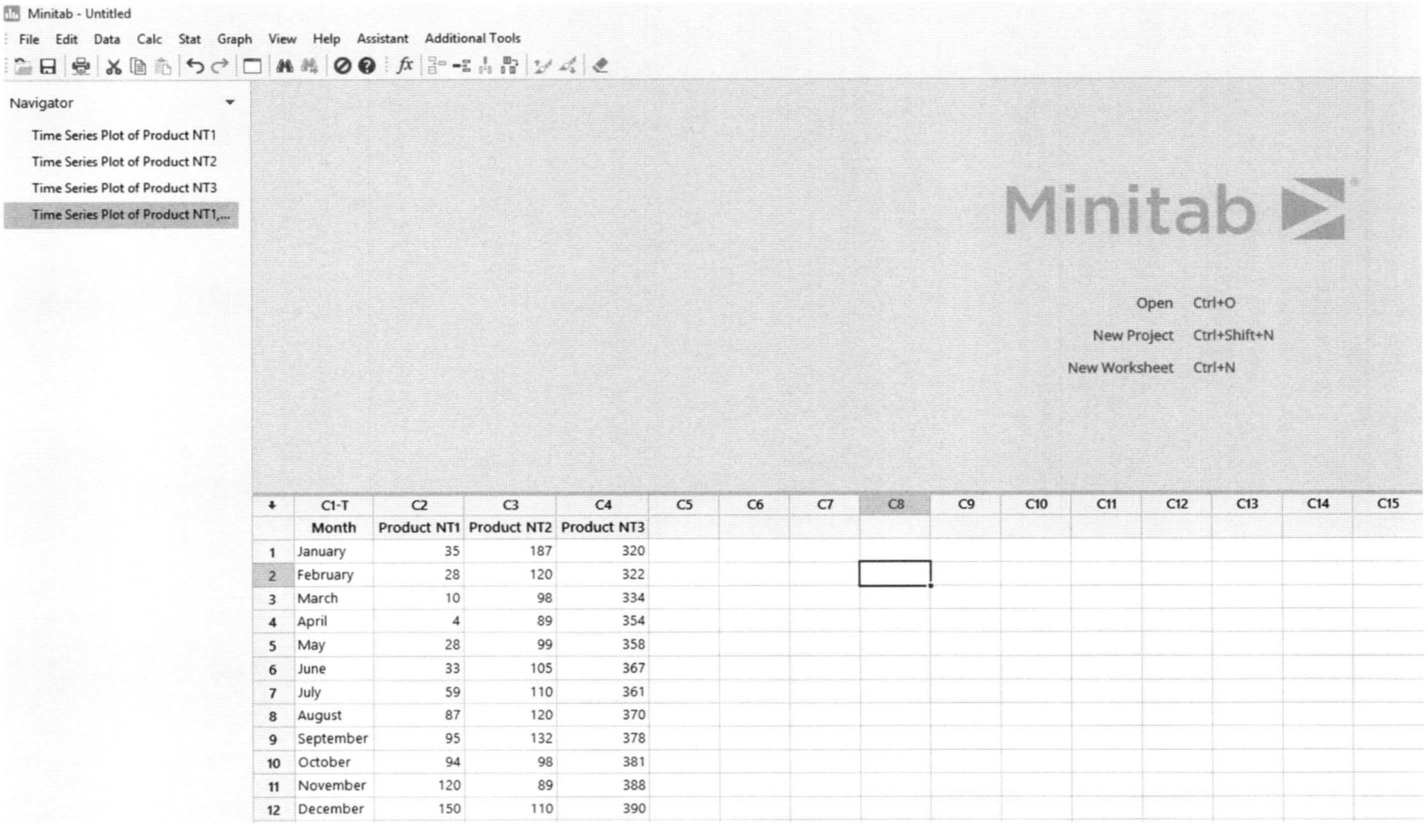

Minitab - Untitled

File Edit Data Calc Stat Graph View Help Assistant Additional Tools

Navigator

Time Series Plot of Product NT1
Time Series Plot of Product NT2
Time Series Plot of Product NT3
Time Series Plot of Product NT1,...

Minitab

Open Ctrl+O
New Project Ctrl+Shift+N
New Worksheet Ctrl+N

	C1-T	C2	C3	C4	C5	C6	C7	C8	C9	C10	C11	C12	C13	C14	C15
	Month	Product NT1	Product NT2	Product NT3											
1	January	35	187	320											
2	February	28	120	322											
3	March	10	98	334											
4	April	4	89	354											
5	May	28	99	358											
6	June	33	105	367											
7	July	59	110	361											
8	August	87	120	370											
9	September	95	132	378											
10	October	94	98	381											
11	November	120	89	388											
12	December	150	110	390											

Figure 5.12 Minitab data visualization – line graph for product NT1

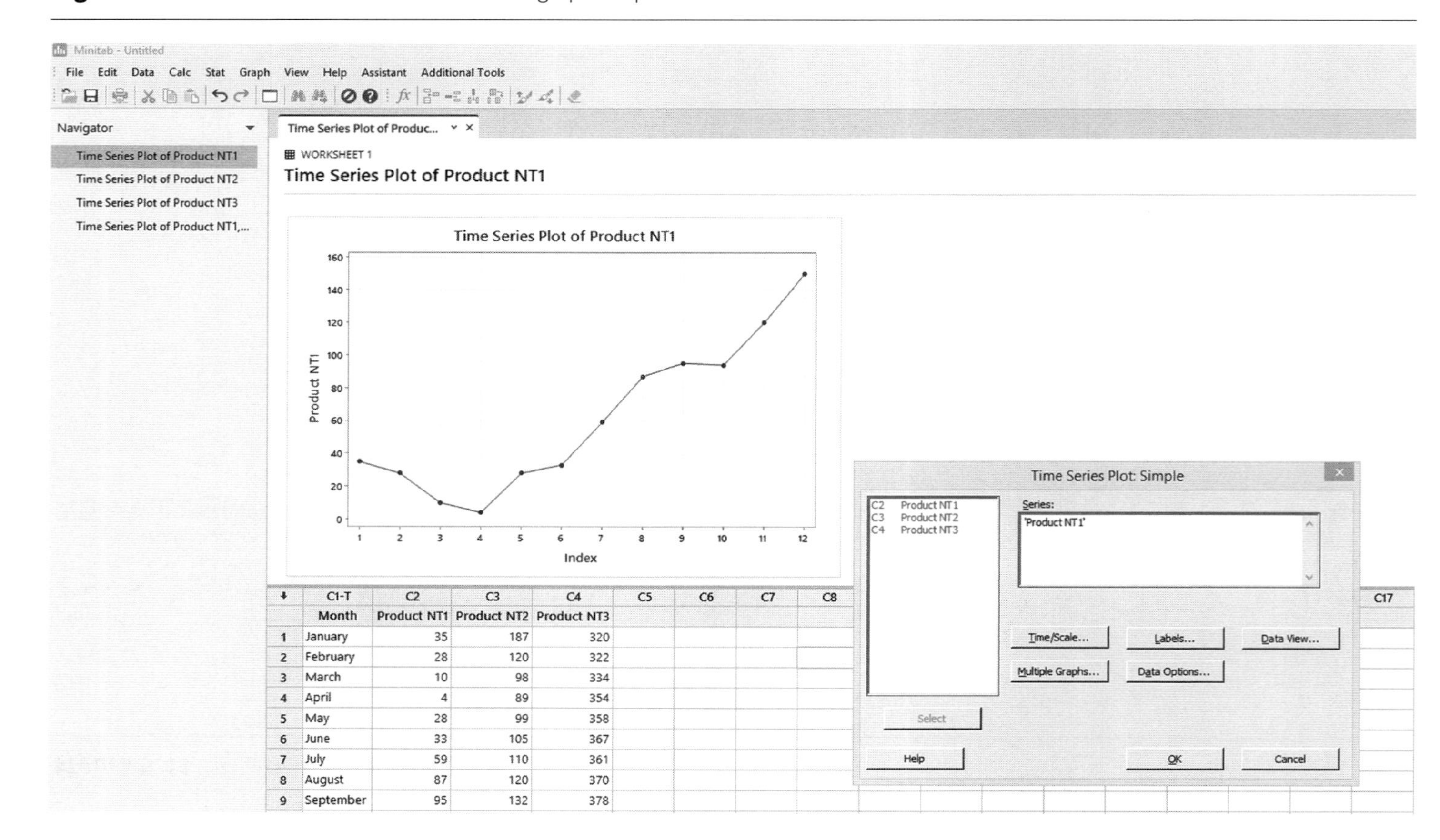

Figure 5.13 Minitab data visualization – all products

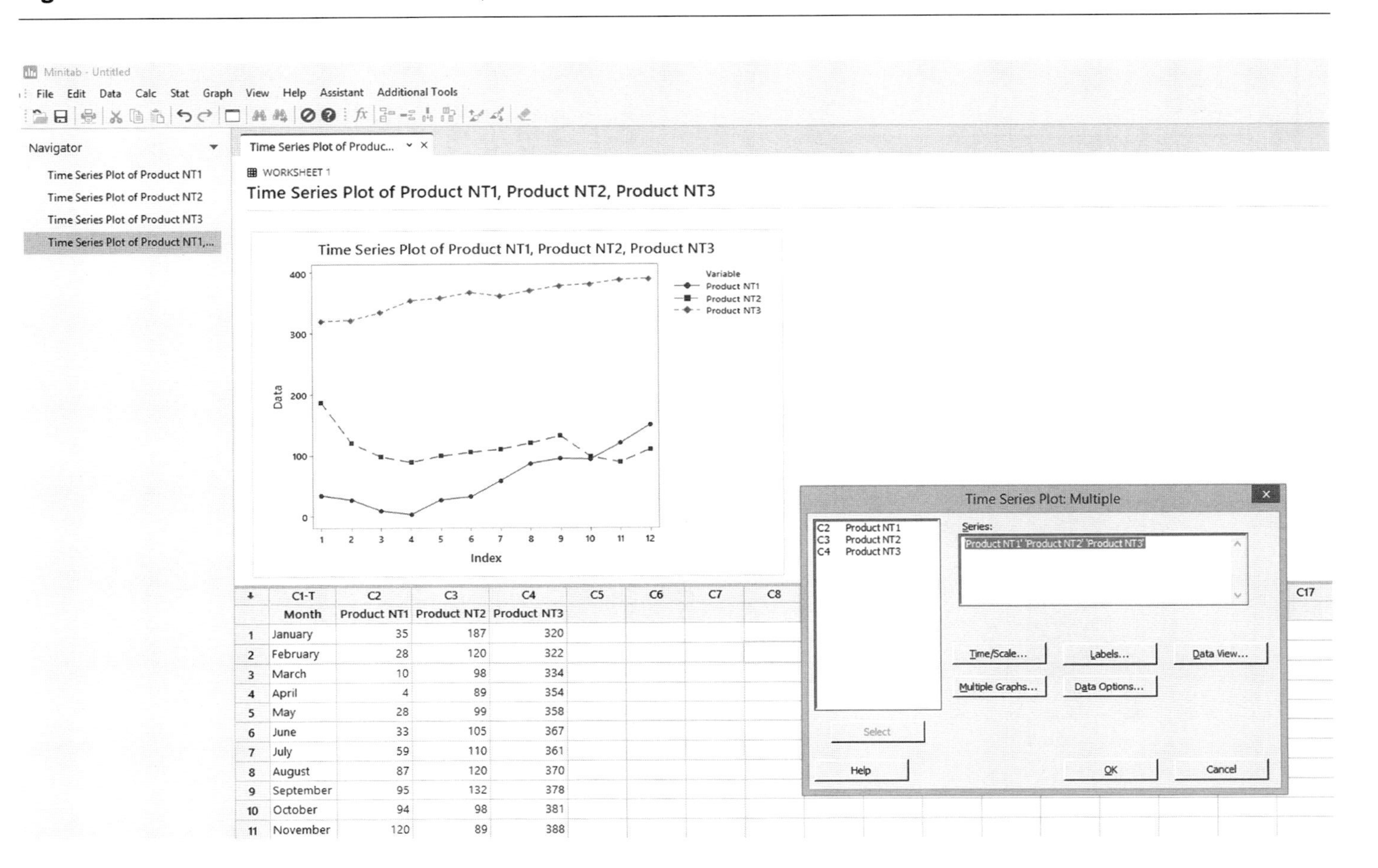

Figure 5.14 Minitab data visualization – all products area graph

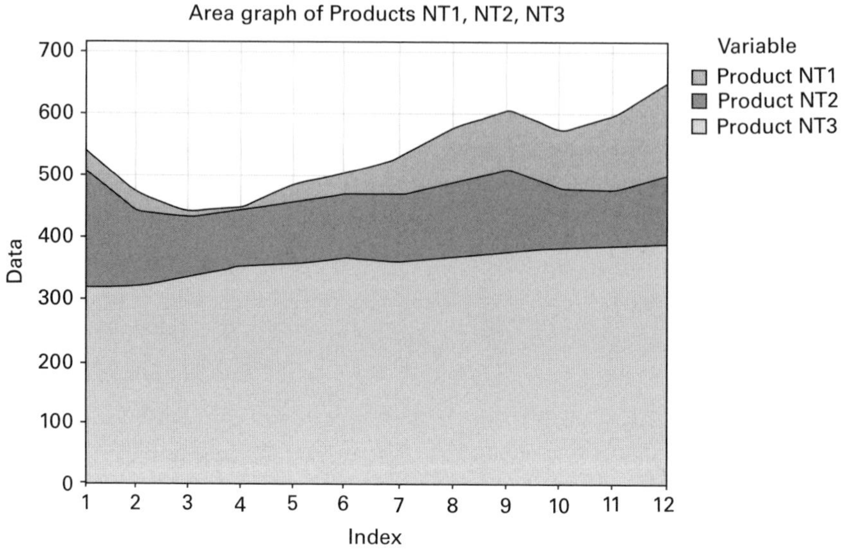

Software packages for data visualization

It has been shown that the manual representation, hand-drawn data shown in Figure 5.1 can also be represented using different computer software packages with, for example, functions in Excel and Minitab. Other programmes that can be used to visualize data are Statistical Analytical System (SAS), R (an integrated development environment that offers a free user-friendly platform for analysing data (Gupta, 2018)), Python, SPSS, Microsoft Power BI and Tableau, to name just a few. On top of these, Sahay (2017), also mentions the visualization tools SOFA, D3 and JavaScript. These software packages support the representation of quantitative data; however, data that is in a qualitative format can also be represented using software packages such as NVivo.

From the graphs or charts presented so far, it can be noted that a variety of them can be used, and they are offered within a number of different software packages, though there is no one single graph that should be used above others. The decision on which graph to use should be based on the data that is available and the task at hand. However, in any situation, from the selected graphs available, it is best to select the one that is the easiest to understand.

Figure 5.15 Minitab data visualization – all products bar chart

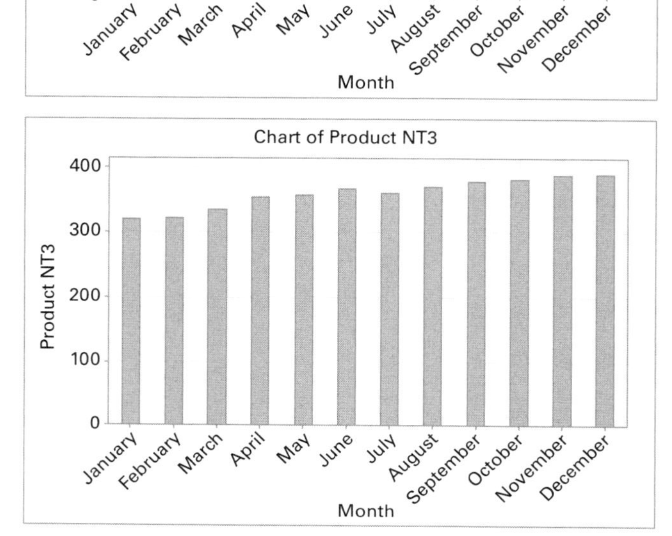

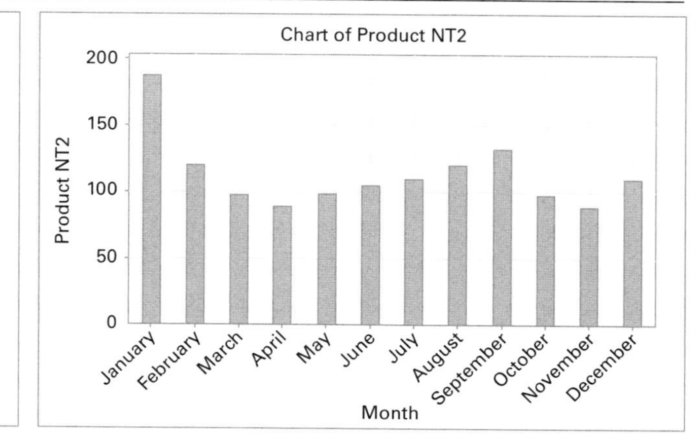

Figure 5.16 Minitab data visualization – bar chart all products comparison

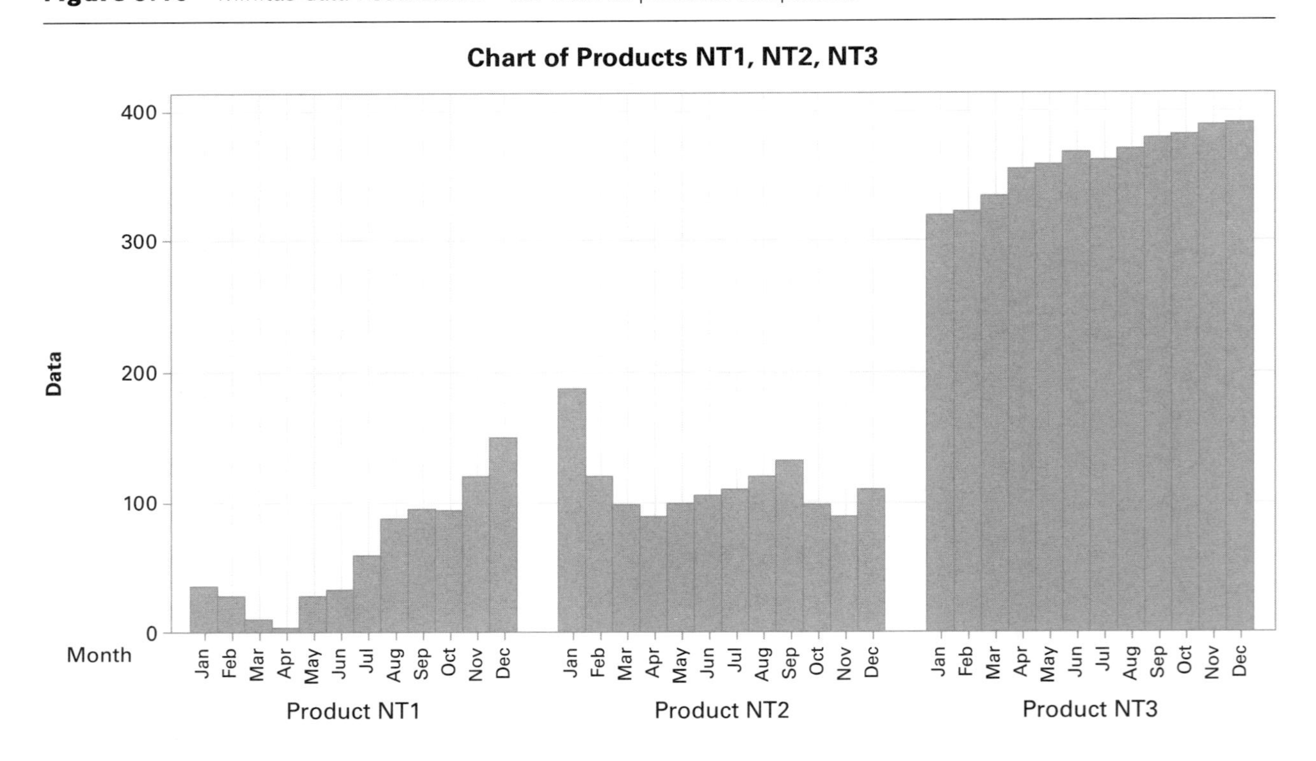

Dashboards are intended to provide a visual display of information collected in aggregated or disaggregated format from different systems of business intelligence or performance measurement systems used within organizations to present the most sought-after information.

Process, information and network visualization

Process and information visualization uses process charts, flow charts, value stream mapping, and cause-and-effect diagrams to represent information (Sahay, 2017). These tools have been used to understand the flow of operations, to evaluate the logical sequence of activities, or to map particular tasks of activities. They have also been used to study operations of, for example, facility location designs and to evaluate the opportunity for implementation and improvement. In supply chain operations these tools, also known as quality tools, are predominantly used by supply chain and operations managers, but not only; they have also been used in finance, manufacturing, different engineering functions representing processes and activities, in process and product design, and many other fields.

Process flow diagrams

Where the intention is to study a process for the purpose of improvement, a number of visualization tools have been developed to ensure this is possible. A very simple process can be represented as in Figure 5.17, where the diagram indicates the input, the process and output. In this case, processes may be seen as an activity, or a set of activities, that receives the input (this can be as data or as other activities), takes this input and transforms this into output. The output can be in the form of data at the end of the process, or can be seen as input to the next process or set of processes. Data collection points in this case can be represented before the process, during the process as work in progress and after the process as the process operations being completed. The process flow diagram may have this basic representation, or it may be extended to represent a set of linked activities or processes. An example of a basic process flow diagram with a number of processes linked in series is represented in Figure 5.18. Each process takes an input and transforms this into an output, then the output from Process 1 is seen as input to Process 2 and so on. Data can be collected and visualized at the beginning, before Process 1 starts and at the end after Process 4 has been completed. Or for a more detailed analysis,

Figure 5.17 A basic process flow diagram

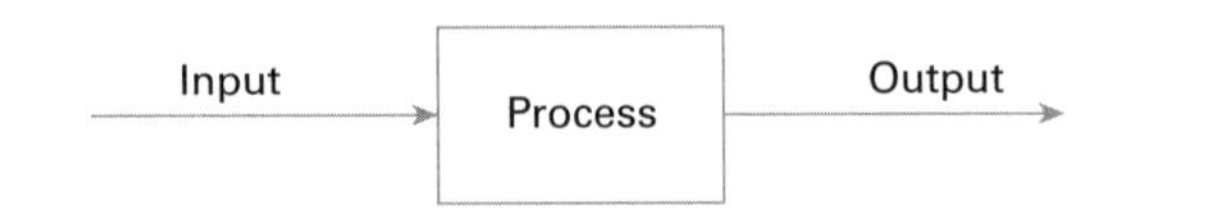

Figure 5.18 A basic process flow diagram with multiple processes connected in series

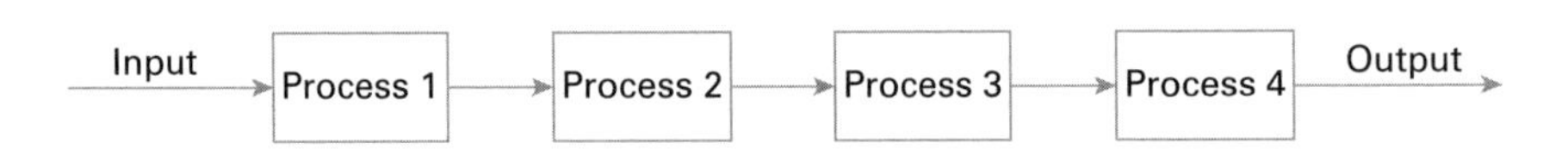

data can be collected before Process 1, at the end of Process 2, Process 3 as well as Process 4. Process specific data can also be collected with details about the work in progress of each activity, still this is also captured as an aggregated value in the output data of each individual stage.

Process flow diagrams may have other representations, such as the one in Figure 5.19, where a number of processes are connected based on different set decisions. Data can be collected at different points in the system, with a sample represented in Figure 5.20, where more data collection points can be set for the example in Figure 5.19.

Processes can be visualized as part of a system, and processes may be formed from a set of sub-processes. The sub-processes may also be formed by a number of smaller individual tasks. A visual representation of the links between these can be a particular help in understanding the bigger picture of where individual activities are taking place.

Tree diagram

A similar representation to the one in Figure 5.20 is the tree diagram, shown in Figure 5.21. Hierarchical processes can be represented using this diagram. This type of diagram is commonly seen in the analytical hierarchical process (AHP) method.

Feedback control diagram

A feedback control diagram can be represented as in Figure 5.22. This is intended to indicate that information is being generated as output that can be considered to control the input of the system. An example of this representation in the supply chain is when the feedback received from customers is considered as part of the input data to improve the process.

Figure 5.19 An example of a flow diagram

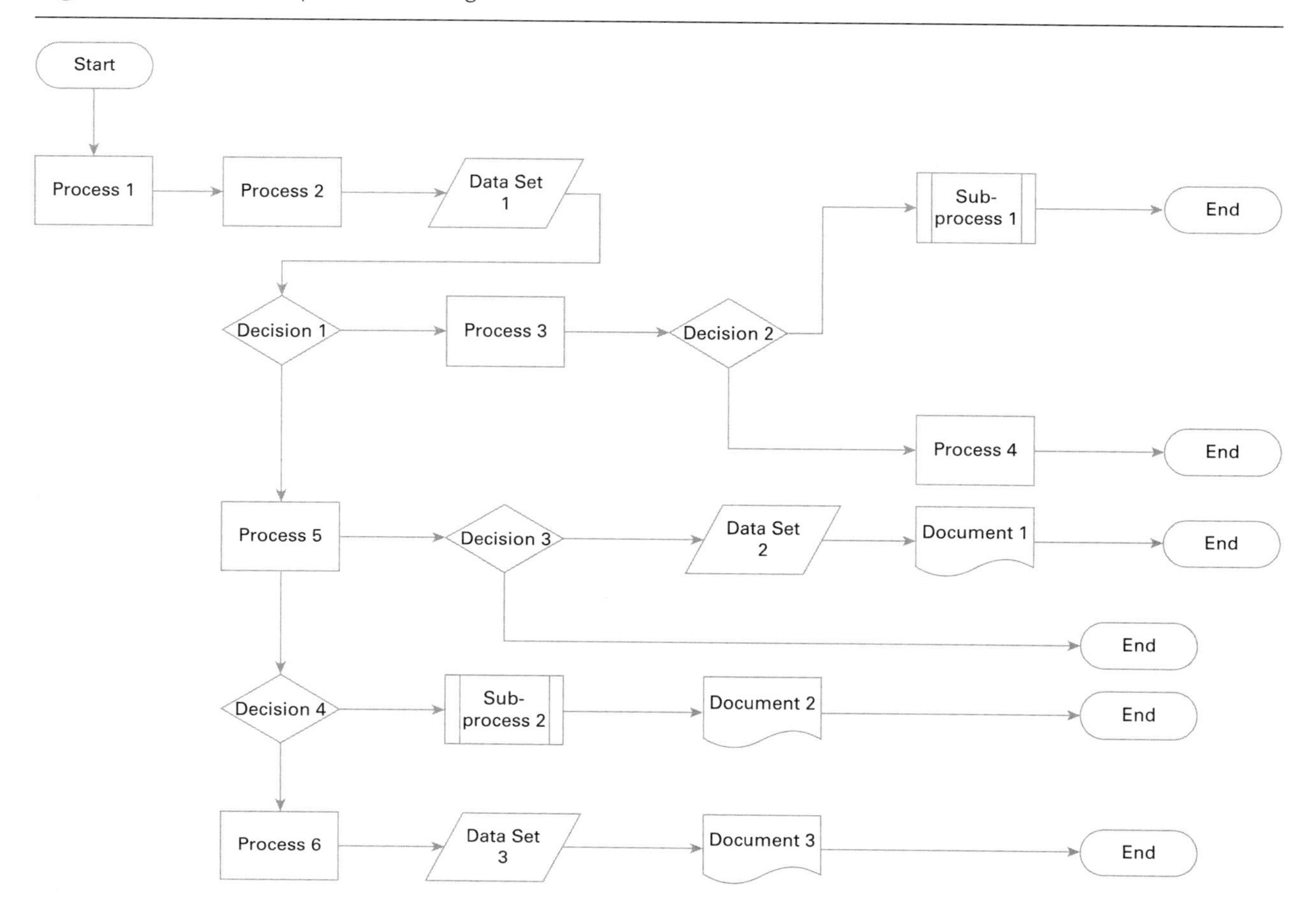

Figure 5.20 A system representation of processes and sub-processes

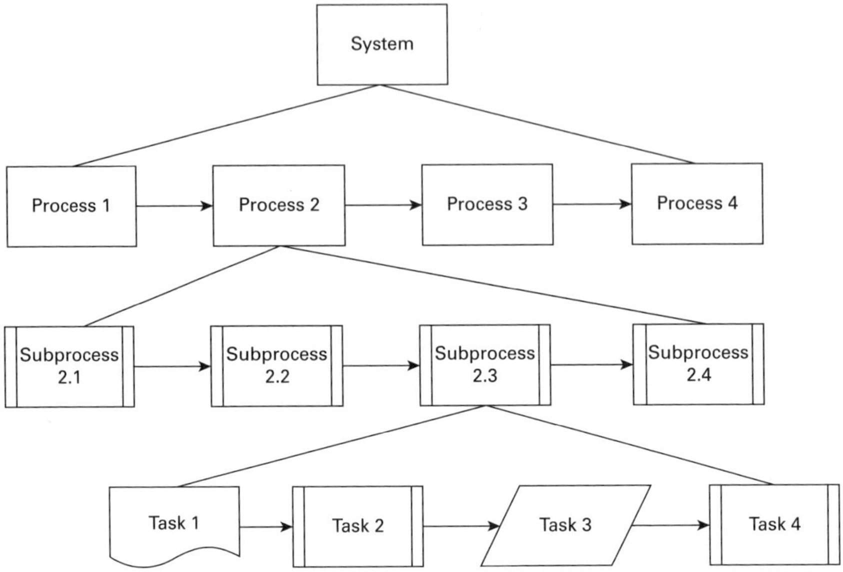

System

Process 1 → Process 2 → Process 3 → Process 4

Subprocess 2.1 → Subprocess 2.2 → Subprocess 2.3 → Subprocess 2.4

Task 1 → Task 2 → Task 3 → Task 4

Value stream maps

In a supply chain, a value stream map (VSM) is used to visualize the flow of activities that are taking place for a particular product or a service. They are intended to review the flow of information and inventory carried from the point of origin (supplies) until the material is changed into a finished product and reaches the customer (see Figure 5.23). A number of details are captured in this diagram, used for analysis, and the non-value-adding activities are being identified with the scope to be eliminated. Value stream maps are used to identify bottlenecks in the supply chain and help the decision-maker to eliminate waste. Value stream maps are not only used in the supply chains, they are also used in other fields, such as manufacturing, software engineering and healthcare. This type of diagram can be represented manually, or can be designed using computer software such as Visio.

Affinity diagram

This type of diagram is used very frequently in many situations when ideas are being generated initially as a brainstorming exercise, after which these ideas are grouped together to form themes. This tool is mainly used to 'organize information and help achieve order out of chaos' (Kubiak and Benbow, 2016, p 181). Using this approach to analyse data, decision-makers can benefit from grouping a large amount of information, concepts and

Figure 5.21 A representation of a tree diagram in the form of an analytical hierarchy process framework

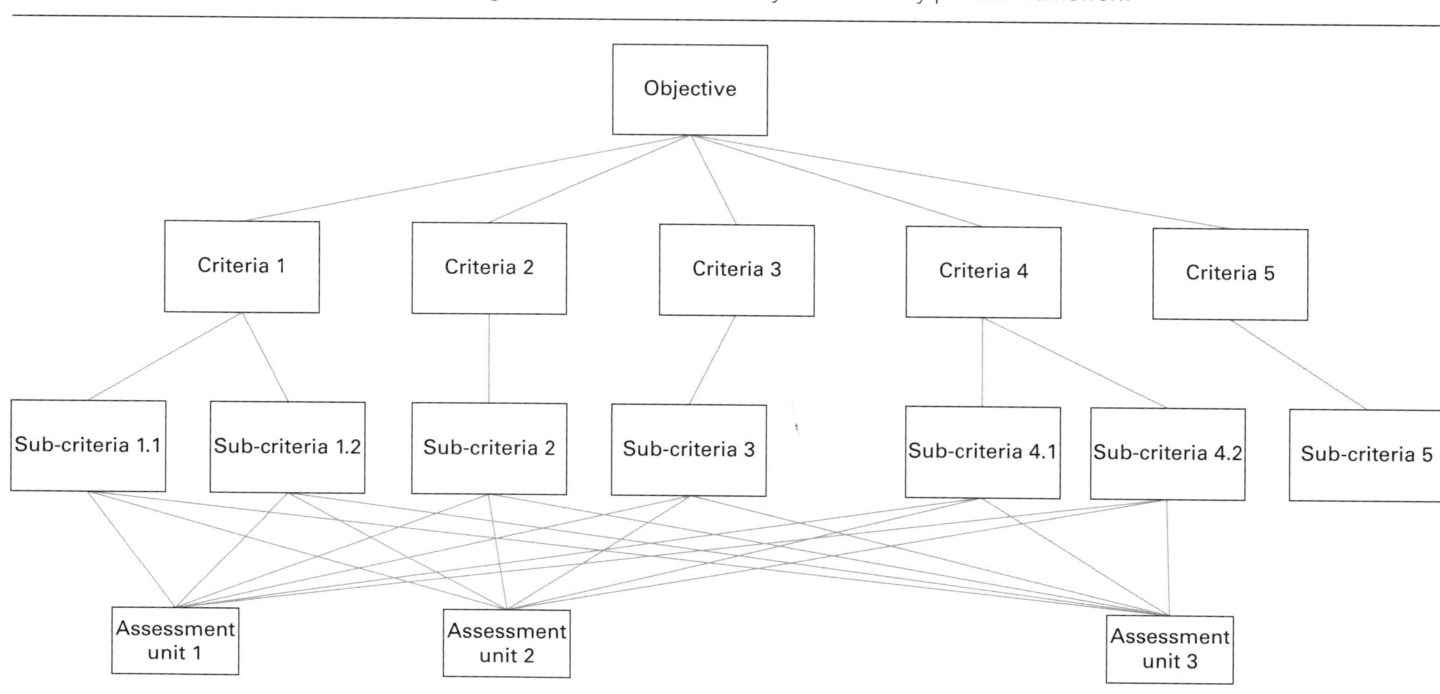

Figure 5.22 A feedback control diagram

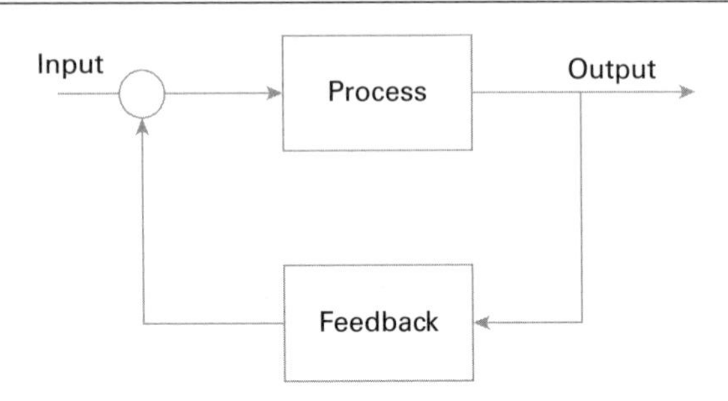

ideas. To construct this diagram, initially ideas based around a specified topic are being generated and pinned to a board, or an electronic board. Following this step, themes are being identified by participants, and original notes belonging to particular themes are being grouped together in a logical way (see Figure 5.24). Participants may consider that a note belongs to more than one theme, and in this case a duplicate of a particular note can be considered. These types of diagrams are heavily used and are particularly valued, as ideas are being generated without a predefined agenda.

Supply chain network view

Following from the concepts presented in Chapter 3, as supply chains can be represented as systems as well as networks, visualizing their structure and further enhancing the way supply chain systems are observed and interpreted are key to modelling and analysing them.

A supply chain may be analysed following a *dyadic* representation (see Figure 5.25). Within this type of representation, the analysis is carried out between two entities that form the system of analysis. They can for example be an analysis between a supplier and a manufacturer, or between a manufacturer and a retailer, or between two retailers, and so on. The link between the two elements can be according to the product flow, or information flow that is bidirectional.

A supply chain may also be evaluated as a *triad*, where a typical triadic representation can be considered (see Figure 5.26). This may appear in a distribution setting, when a regional warehouse distributes products to two local retailers (Figure 5.26 a), or it can appear in an analysis of supplier selection, when the manufacturer is deciding between which of two suppliers to select (Figure 5.26 b). In any of these combinations, only three elements are considered for analysis. Figure 5.26 indicates a representation of a triad with two links.

Figure 5.23 A template for a value stream map

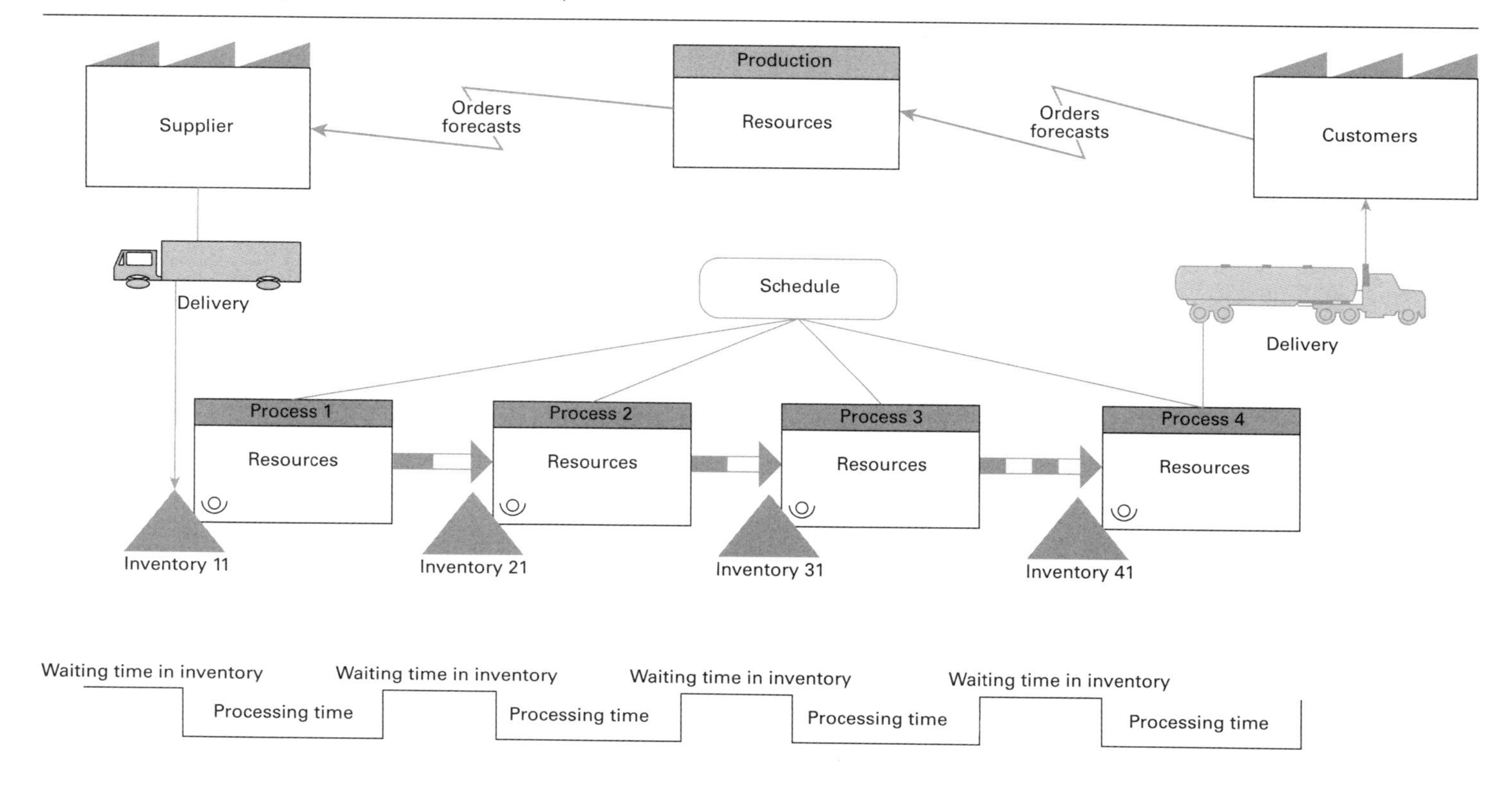

Figure 5.24 An example of an affinity diagram

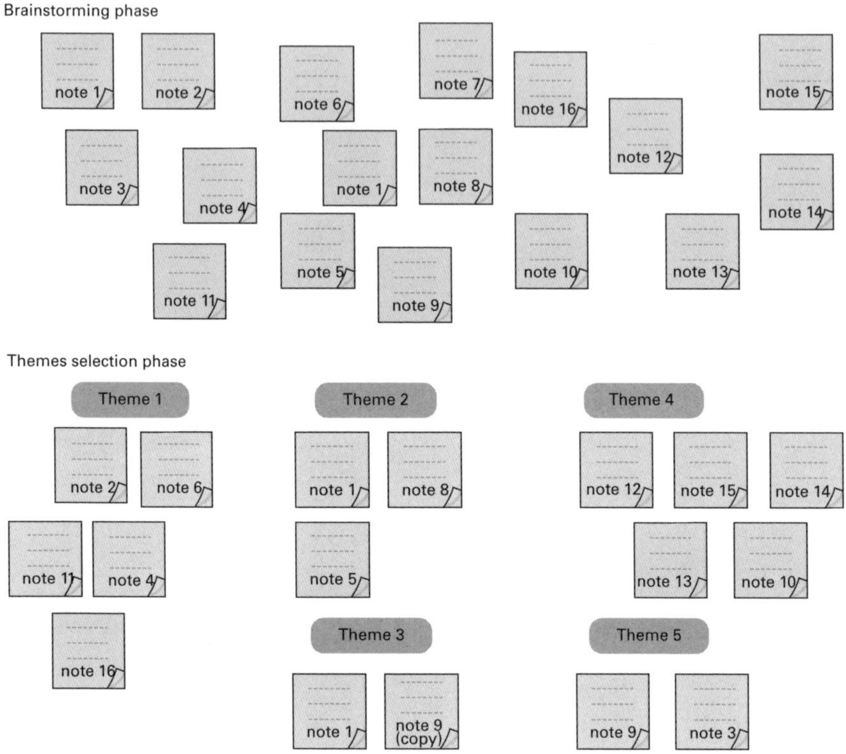

Following this concept, supply chains have been represented using divergent and convergent representations (Beamon and Chen, 2001). A *divergent* representation of a supply chain is typical in a distribution problem (see Figure 5.27 a). The links indicated in this case are typical in representing the flow of goods in the chain; however, the information link may be indicated with different connections in the chain, depending on the level of visibility the supply chain has with its partners. Within a *convergent* representation of a supply chain, the analysis is more concerned with aggregation, or assembly (see Figure 5.27 b), where it may be that, from a number of suppliers and suppliers' suppliers, the analysis is focused on one output at, for example, a manufacturing location.

As soon as the analysis considers the extended chain forming the link between the focal company in the middle with their suppliers and their suppliers' suppliers, and customers and their customers' customer, the visual representation will be in a *conjoint* format (see Figure 5.28). Similar representation of supply chains has been noted in Beamon and Chen (2001), Lambert et al (1998) and Lambert and Pohlen (2001), where they refer to these representations as supply chain networks, or 'inter-company, business process links'.

Figure 5.25 Dyadic representation of a supply chain

Figure 5.26 Triadic representation of a supply chain

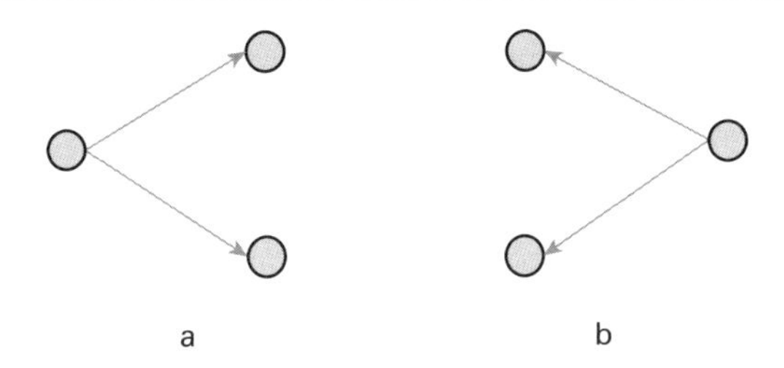

Figure 5.27 Convergent and divergent representation of a supply chain

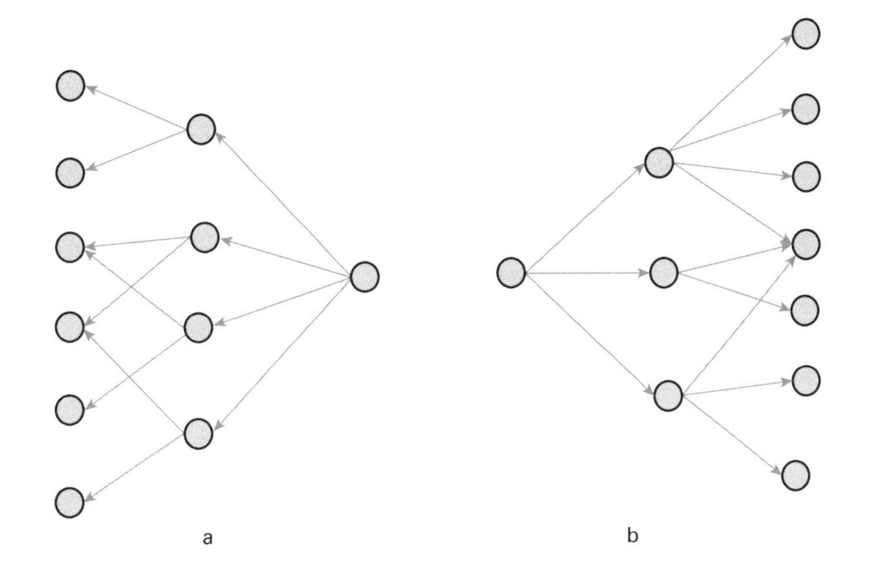

The supply chain can be represented as a *network*, where the analysis is carried out based on the entire structure, or predefined elements within a given structure (see Figure 5.29). Many variations of this representation of supply chain networks have been presented in the literature, where each aims to indicate the complex nature of the supply chain network structure. One other characteristics of these representations is the fact that there is no predefined focal company, and the analysis is intended to meet the scope of the entire supply chain. Still, these representations do not necessarily capture more than the physical flow in the supply chain. It is relevant to mention that information, ordering and so on can also be represented in this case.

Figure 5.28 Conjoint representation of a supply chain

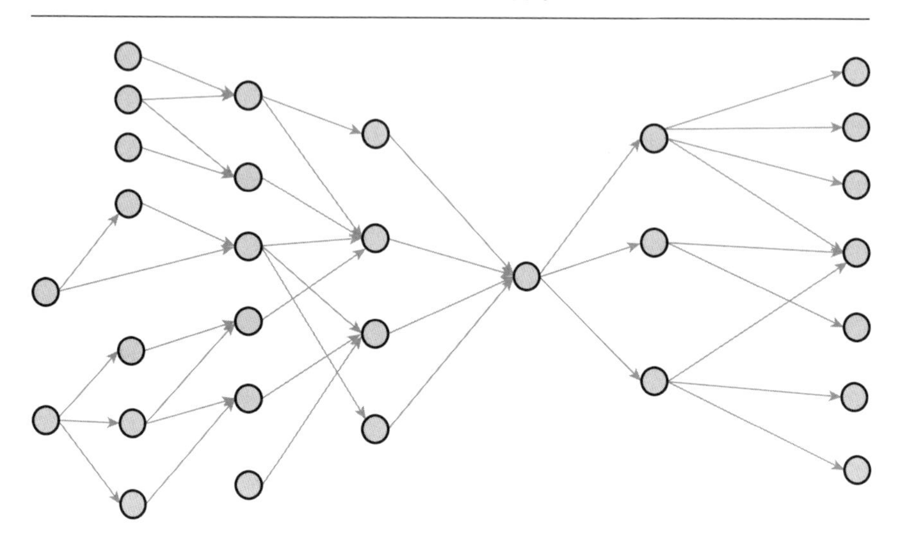

Figure 5.29 A supply chain network structure

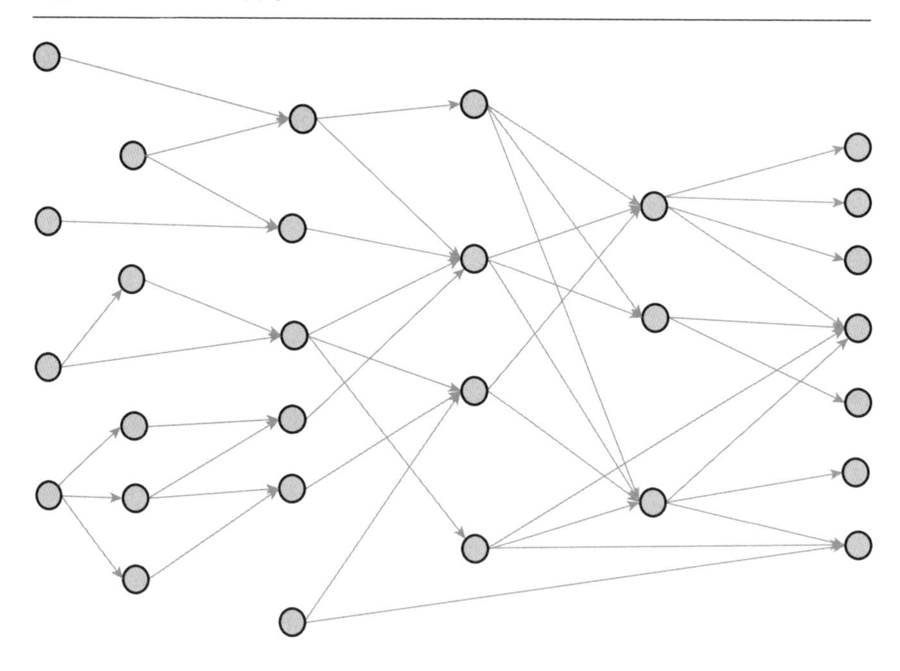

Summary

This chapter aimed to present the visual tools that can be used to represent data, processes, information and systems in the supply chain. A number of analytical techniques can be used to represent data in the supply chain, and a number of graphs and visualization tools have been and can be used to represent, evaluate and control data. From Chapters 3 and 4 we also see the various formats data can take and the collection points typical for supply chain systems.

It is particular important to understand how data can be represented, as well as how processes can be visualized to help in the decision-making process. A number of process diagrams have been displayed in this chapter. Processes are smaller representations of systems and networks, and for supply chain systems, understanding the visual representation of networks is particularly important.

A number of data and process flow diagrams have been demonstrated in this chapter; however, in the area of supply chain many more representations have been used by authors and practitioners over the years. Many of these diagrams can be represented manually, but there are a number of software packages that can be used to insert or draw these diagrams. Using these tools, visual boards can be designed to represent a set of performance measures belonging to the data or processes being analysed.

References

Beamon, B M and Chen, V C P (2001) Performance analysis of conjoined supply chains, *International Journal of Production Research*, 39(14), 3195–218

Gupta, D (2018) *Applied Analytics Through Case Studies Using SAS and R: Implementing predictive models and machine learning techniques*, Apress, Berkeley, CA, doi:10.1007/978-1-4842-3525-6

Kubiak, T M and Benbow, D W (2016) *The Certified Six Sigma Black Belt Handbook*, 3rd edn, ASQ Quality Press, Milwaukee, WI

Lambert, D M and Pohlen, T L (2001) Supply chain metrics, *The International Journal of Logistics Management*, 12(1), 1–19

Lambert, D M, Cooper, M C and Pagh, J D (1998) Supply chain management: Implementation issues and research opportunities, *The International Journal of Logistics Management*, 9(2), 1–19

Sahay, A (2017) *Data Visualization: Volume II – Uncovering the hidden pattern in data using basic and new quality tools*, Business Expert Press, New York

Bibliography

Bendoly, E (2016) Fit, bias, and enacted sense making in data visualization: Frameworks for continuous development in operations and supply chain management analytics, *Journal of Business Logistics*, 37(1), 6–17, doi:10.1111/jbl.12113

Sosulski, K (2019) *Data Visualization Made Simple: Insights into becoming visual*, Routledge, New York

Business analytics 06

Descriptive and predictive models

LEARNING OBJECTIVES

- Develop an awareness of typical descriptive and predictive models used in the area of business analytics.
- Understand a set of descriptive BA models typically used in the area of supply chain.
- Understand a set of predictive BA models typically used in the area of supply chain.
- Learn the approach on working with BA models and discuss potential results.
- Understand how to develop and analyse descriptive and predictive models in Excel and Minitab.

Introduction

This chapter mainly focuses on discussing a range of business analytical aspects and providing a number of descriptive and predictive analytical examples. These examples have applicability in different aspects of business and management, however the issues discussed within these examples are approached from a supply chain angle. The models introduced here are critically discussed where the issues these models can solve are presented. The way in which these models can be solved is presented where the challenges linked to their analysis and implementation are also considered.

The models presented are grouped into two distinct categories: descriptive models and predictive models.

As indicated in the previous chapters of this book, a number of commercial software packages are available and can be used to support the application of many of these models. Some of these software packages may have the capability to solve only one aspect of the problem, where others may allow for a more integrated approach. Some of these software packages may require advanced training for the analysts and users to create and operate the models created, which incurs additional costs.

In many situations, the models discussed in this section can be implemented when using spreadsheet software. Some of the examples developed in this chapter use Microsoft Excel 2016 for Windows and Minitab software. It is assumed that the reader is familiar with some of the basic functions in Excel, however the functions used and the way these have been considered are detailed in each case.

Descriptive models

As discussed in earlier chapters, descriptive models are those looking to answer questions on what happened, why it happened and what is happening now. Data visualization models are critical in supporting the decision-making process and interpreting results. Visualizing data, and using descriptive analytics to understand the relationship between different variables in the data, will help an analyst, for example, to select the most appropriate forecasting model for prediction. Frequency distribution and histograms form part of the descriptive statistical models, however as these are visual tools, some descriptive models have been presented in the previous chapter (see Chapter 5).

Descriptive models are characterized by the use of statistical analysis. A number of statistical models can be used in the field of business and management with particular application to the supply chain. Some of these are related in this section, and more examples can be found in Curwin and Slater (2008, 2004), Curwin et al (2013), Evans (2016), Kubiak and Benbow (2016) and Field (2018), to name just a few books in this field.

In many instances, statistical analyses are carried out on a sample of data, as it may ether be impossible to evaluate the entire data (this is known in statistical terms as the *population*), or data may not be available or not appropriately recorded.

A number of notations are being considered when representing statistical models. Notations of data are being used such as x_1, x_2 and so on, where this array of values can also be represented as x_i, where *i* indicates the number of observations.

When representing an average, the following representation can be used:

$$\bar{x} = \frac{\sum_{i=1}^{n} x_i}{n}$$

Where $\bar{x}$ represents the notation for the mean, and *n* the total number of observations. The function to be used in Excel will be =AVERAGE (data range).

For example, if the data for the total number of products produced in a week is captured and listed as indicated in Figure 6.1, the average per week for the 20 weeks of data (20 observations) will be =AVERAGE(B2:B21) with a calculated result of 410.9 and represented in cell B22.

Other descriptive analysis that could take place for the data indicated in Figure 6.1 are the total number of products produced in 20 weeks, using the formula = SUM(B2:B21). Other functions can be used, such as COUNT, which counts the number of observations. For this particular example, we know the data from the number of weeks that is also represented in column A; however, this will be another way to count the number of observations used in calculation. Functions such as MAX and MIN, where MAX indicates the maximum vale produced and MIN indicates the minimum vale produced, can also be considered and will show the extreme values obtained during the period analysed. Other observations that could be evaluated in this case are the median, mode and standard deviation, where their associated Excel functions used are presented in Figure 6.1.

This data can also be visualized by inserting a scatter chart with smooth lines.

The analysis can continue by adding a linear trendline to the data collected. One other option available in Excel is to add a trendline option such as a three-point moving average (see Figure 6.2). This approach will continue to be further explored in the section on predictive models, and the identified function of a trendline could also be used for prediction.

Similar analyses can be carried out in Minitab. Data from Excel can be copied in Minitab, in the spreadsheet view (see Figure 6.3).

Select *Stat – Basic Statistics – Display Descriptive Statistics* and the results are obtained as presented in Figure 6.3.

Figure 6.1 Descriptive analysis for a weekly production

	A	B	C
1	Weeks	Total number produced	*Formulas used*
2	1	350	
3	2	275	
4	3	451	
5	4	321	
6	5	470	
7	6	450	
8	7	420	
9	8	410	
10	9	432	
11	10	398	
12	11	399	
13	12	401	
14	13	430	
15	14	435	
16	15	380	
17	16	468	
18	17	410	
19	18	400	
20	19	468	
21	20	450	
22	Averege/week	410.9	=AVERAGE(B2:B21)
23	Total produced	8218	=SUM(B2:B21)
24	No of observations	20	=COUNT(B2:B21)
25	Max value	470	=MAX(B2:B21)
26	Min value	275	=MIN(B2:B21)
27	Median	415	=MEDIAN(B2:B21)
28	Mode	450	=MODE.MULT(B2:B21)
29	Standard deviation	48.9355699	=STDEV.P(B2:B21)

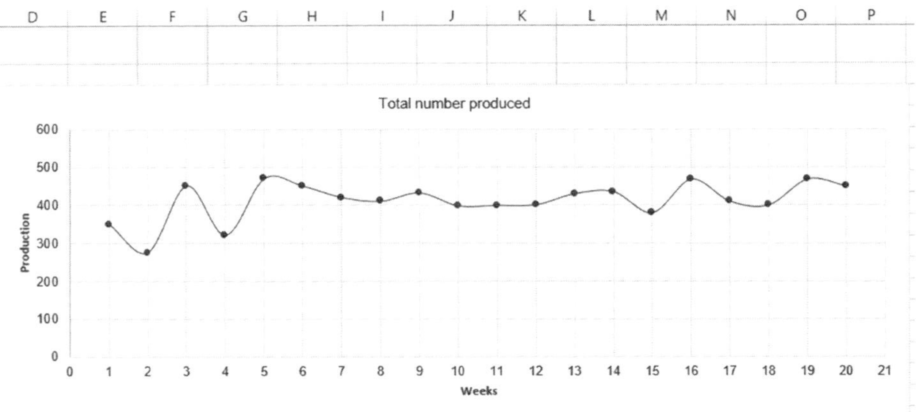

Figure 6.2 Descriptive analysis visualizing production data

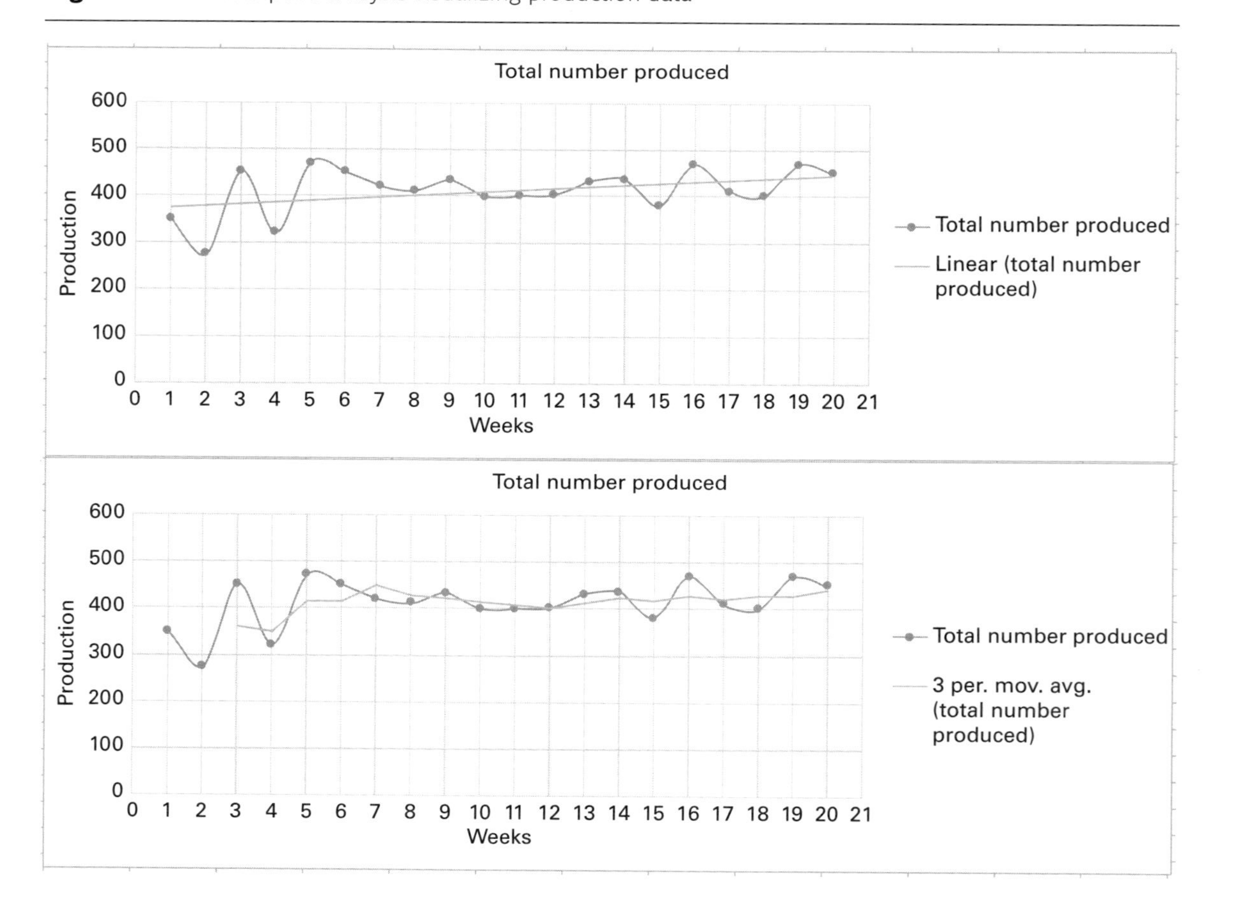

Figure 6.3 Descriptive analysis in Minitab

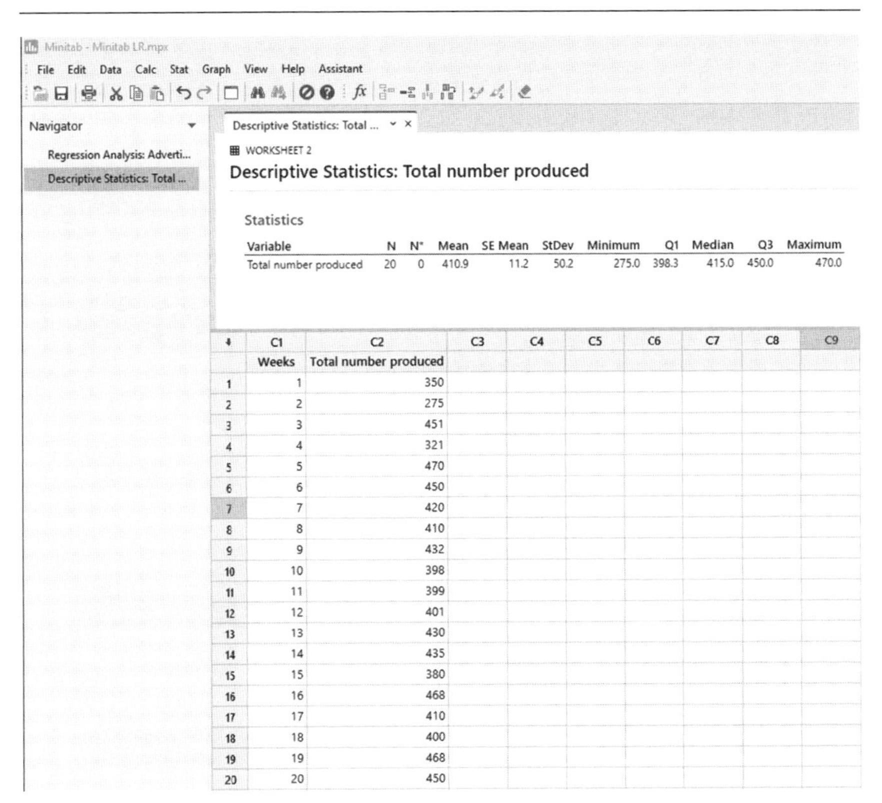

Variable	N	N*	Mean	SE Mean	StDev	Minimum	Q1	Median	Q3	Maximum
Total number produced	20	0	410.9	11.2	50.2	275.0	398.3	415.0	450.0	470.0

	C1	C2	C3	C4	C5	C6	C7	C8	C9
	Weeks	Total number produced							
1	1	350							
2	2	275							
3	3	451							
4	4	321							
5	5	470							
6	6	450							
7	7	420							
8	8	410							
9	9	432							
10	10	398							
11	11	399							
12	12	401							
13	13	430							
14	14	435							
15	15	380							
16	16	468							
17	17	410							
18	18	400							
19	19	468							
20	20	450							

Histograms are used when we are interested in understanding the distribution of numerical data. The first step into building a histogram is to collect the data that will be analysed. For example, in Figure 6.4 the total number of items sold for 20 products is collected. Following this step, we will need to distribute the range of values for the data available for analysis into intervals. These intervals are also called 'bins' of data (see Figure 6.4, column D). In this particular example six bins have been created from 0 to 20, from 21 to 40, from 41 to 60 and so on. The frequency of having items sold in each defined category can be calculated using the Excel function =frequency (data_array, bins_array) as the cells selected from E2:E7 =FREQUENCY(B2:B21,D2:D7).

The values from the columns D2:E7 will form the data that can be represented in the histogram.

A short list of descriptive models has been included here, and a number of other models are well detailed in many statistical books (Albright and Winston, 2011; Curwin et al, 2013; Field, 2018, and others).

Figure 6.4 Example of a histogram

	A	B	C	D	E
1	**Products**	**Items sold**		**Bin**	**Frequency**
2	Product NT1	43		20	3
3	Product NT2	76		40	4
4	Product NT3	98		60	5
5	Product NT4	32		80	4
6	Product NT5	57		100	3
7	Product NT6	87		120	1
8	Product NT7	28			
9	Product NT8	63			
10	Product NT9	58			
11	Product NT10	48			
12	Product NT11	65			
13	Product NT12	10			
14	Product NT13	7			
15	Product NT14	48			
16	Product NT15	96			
17	Product NT16	114			
18	Product NT17	63			
19	Product NT18	34			
20	Product NT19	27			
21	Product NT20	10			

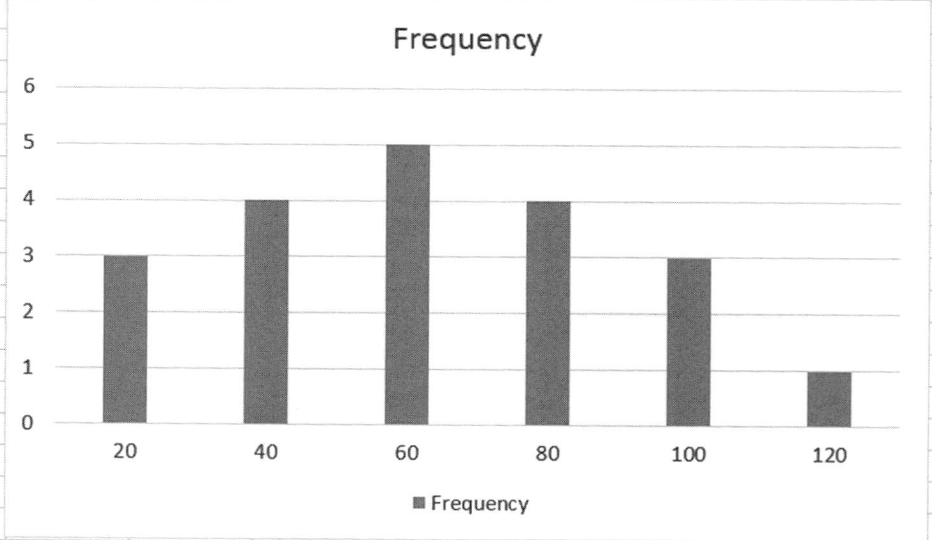

Predictive models

A number of models can be categorized under this section, all of which can be used to support managers in taking the best decision. Lepenioti et al (2020) have conducted a systematic review of the literature looking at prescriptive models identified in the literature from 2010 to 2019. They identify a number of predictive models during this research as well, as some of the predictive models generate results that are also used as input data to prescriptive models. They identify three categories of predictive models used in general (not directly referring to supply chain systems): probabilistic models, machine learning/data mining models, and statistical analyses models. They grouped the models into categories as follows:

- Probablistic: Bayesian network, Markov chain Monte Carlo and hidden Markov model.
- Machine learning/data mining: pattern recognition, random forest, Gaussian process, conditional interface tree, support vector machine, ensemble learning, artificial neural network, random search, decision tree, clustering-based heuristics, k-nearest neighbours algorithm, Kernel methods, multilayer perception and gradient boosted tree.
- Statistical analysis: linear regression, multiple linear regression, rank regression, ARIMA, logistics regression, multinomial logistics regression, density estimation and support vector regression (Lepenioti et al, 2020). Some of these predictive models are also discussed in the section below.

In supply chains, the forecasting function is applied in many parts of the business with the scope to better plan production, predict inventory levels, place orders, optimize transportation and organize the business. A number of forecasting methods have been developed and used within a supply chain and they depend on the type of data they are using, such as data of a quantitative or qualitative nature. A number of methods, such as Delphi technique, executive opinion, sales force pooling and customer services, use qualitative data and are based on opinions and judgements (Merkuryeva et al, 2019). There are also a number of forecasting methods that use data of a quantitative nature and they use past data to predict future data behaviour. Some of these methods are explained later on in this chapter, with methods such as moving average, exponential smoothing methods such as single exponential, double exponential smoothing or Holt's method, and triple exponential smoothing or Holt and Winter's method. There are also methods that look at the relationship between variables, such as regression methods. Within a supply chain, linear regression and multiple regressions (Mumtaz et al, 2018; Merkuryeva et al, 2019), and quadratic regression models (Beamon and Chen, 2001) have been extensively used for prediction.

Other models used for prediction are the time series based models, some of which (moving average, single exponential smoothing, double exponential smoothing, triple exponential smoothing) are detailed in the following section. These are detailed using software packages such as Excel and Minitab. Selecting the best forecasting technique depends very much on the characteristics of the data collected or data available for forecasting.

Regression models

Linear regression models

In many situations in a supply chain, we face the challenge of modelling the relationship between a dependant and an independent variable. For example, we may wish to predict sales based on price, or predict demand based on the promotional activity carried out and so on. For this type of model, a linear relationship can be considered between a dependant variable marked as y and an independent variable marked as x. A linear relationship will take the following form: $y = b_0 + b_1 * x$, where b_0 is the y-intercept and b_1 is the slope of the line.

Regression analyses can be used to evaluate the relationship between a dependant variable and one or more independent variables. When a single independent variable is being considered, the model will take the form of simple linear regression. However, when two or more independent variables are considered in the analysis, this is known as multiple regression. This type of model will be explained in the following section.

The example provided in Figure 6.5 shows the sales figures of a particular product based on the advertising budget allocated for this product. In Excel, these numerical values can be represented using a scatter diagram. Using the Excel *Trendline* tool, a linear model can be selected, as depicted in Figure 6.5. The tool in Excel also allows us to select and display the linear equation that best fits the data and the R-squared that measures the best fit of the line to the data. For the data selected under analysis, the linear equation is: $y = 0.3058x + 6.122$, with the values for $b_0 = 6.122$ and $b_1 = 0.3058$. The value for the R-squared is expected to be between 0 and 1. In this case the $R^2 = 0.974$, where the larger the value of R^2 the better the fit.

Figure 6.5 A linear regression example

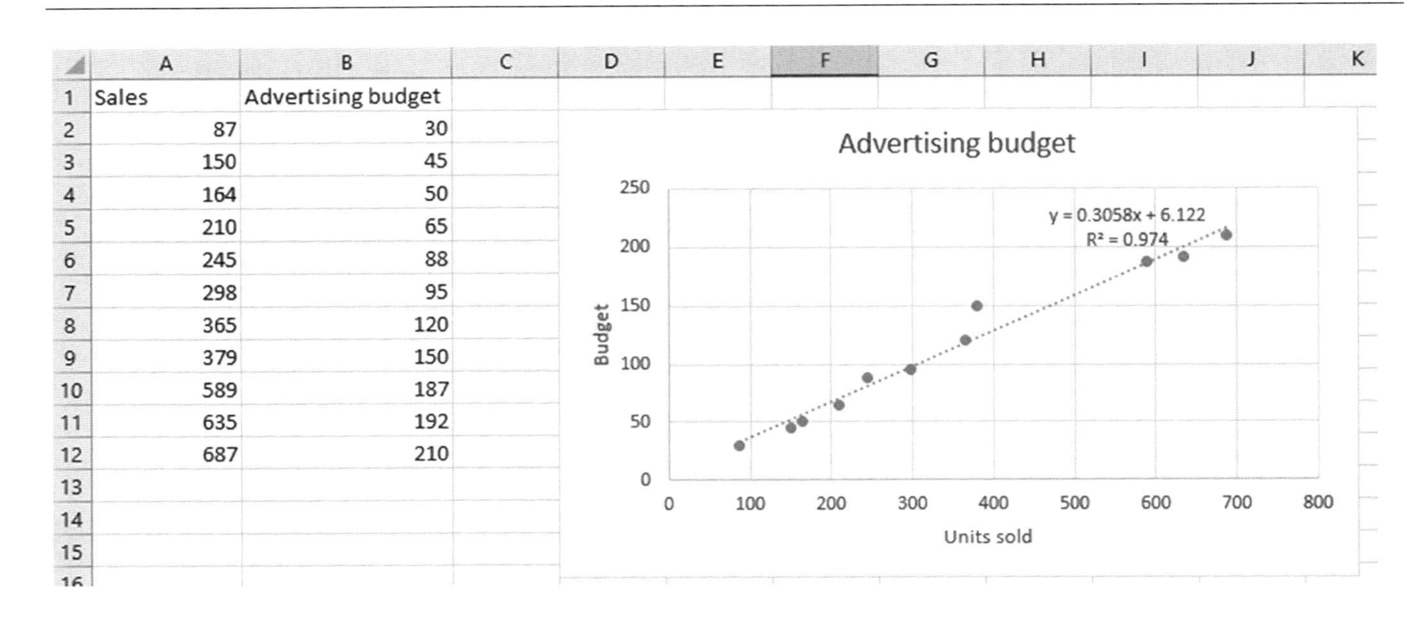

Sales	Advertising budget
87	30
150	45
164	50
210	65
245	88
298	95
365	120
379	150
589	187
635	192
687	210

In Excel, there is a software tool available to carry out regression analysis. This tool can be used for simple linear regression as well as multiple regression.

This tool sits under *Data Tab* and *Data Analysis* under the *Analysis* menu. From the *Data Analysis*, the *Regression* can be selected from the list available (see Figure 6.6).

Figure 6.6 Excel data analysis

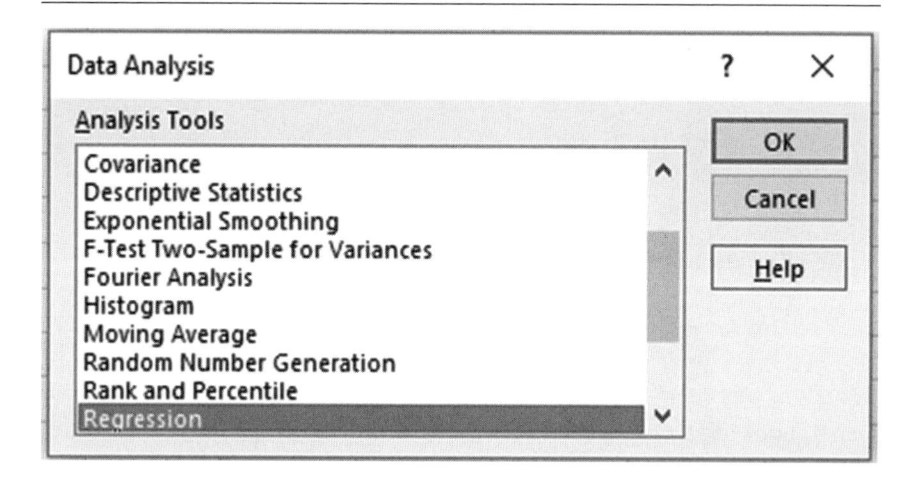

The screen shown in Figure 6.7 indicates the Input Y Range, for which the data under the adverting budget can be considered as $B$1:$B$12. The Input X Range will be the sales data: $A$1:$A$12.

This will result in a basic analysis provided by the Excel Regression tool for the sales-advertising data as indicated in Figure 6.8. The coefficient of determination R squared indicates how much the variation in the dependant variable is considered by the independent variable. The value of R^2 is represented in cell B5, the intercept in cell B17 and the sales coefficient in cell B18. These values are also provided by the *Trendline*, as noted from Figure 6.5.

To be in a position to use this regression analysis for prediction, the model that can be created in this case follows the following formula:

$$Advertising\ budget = 6.12204 + 0.3058 * Sales$$

In other words, for any other value of *sales* figure, we could predict the *advertising budget* using the equation indicated above.

The representation provided above has been all carried out using the data analysis tool available in Excel.

Figure 6.7 Excel data analysis – regression

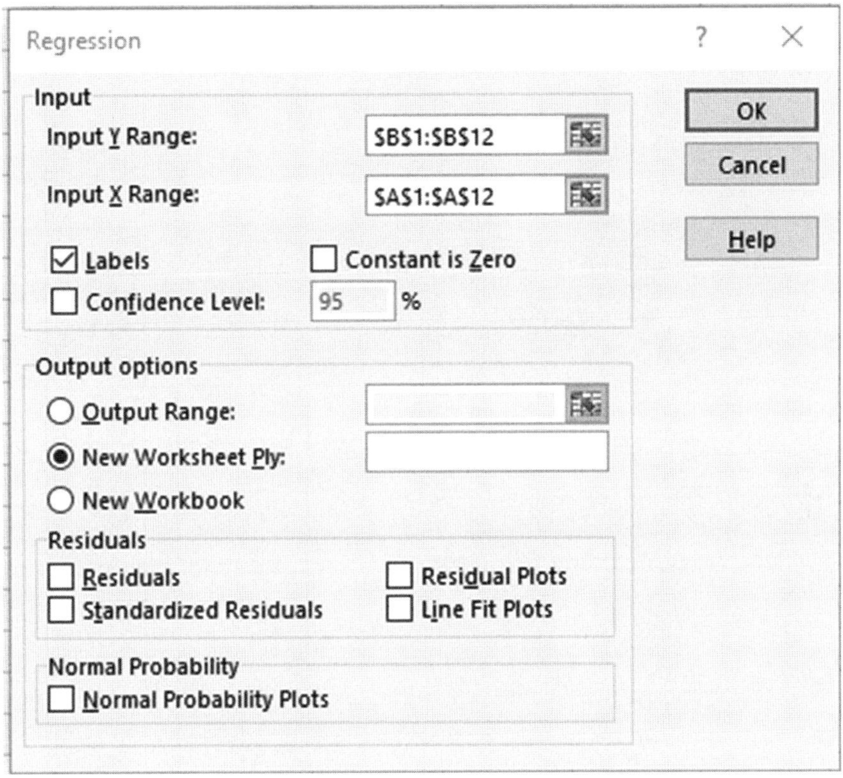

Regression

Input
Input Y Range: $B$1:$B$12
Input X Range: $A$1:$A$12
☑ Labels ☐ Constant is Zero
☐ Confidence Level: 95 %

Output options
○ Output Range:
◉ New Worksheet Ply:
○ New Workbook

Residuals
☐ Residuals ☐ Residual Plots
☐ Standardized Residuals ☐ Line Fit Plots

Normal Probability
☐ Normal Probability Plots

OK
Cancel
Help

The same analysis can be carried out using another statistical software package, for example Minitab. In this case, Minitab 19 is used to demonstrate the way in which a linear regression analysis can be carried out.

The data presented in Excel can be copied into Minitab in the spreadsheet section as indicate in Figure 6.9 under columns C1 and C2.

After the data has been inserted the analysis can start by selecting *Stat – Regression* – and *Fitted Line Plot* for the simple linear regression (see Figure 6.9).

For the *Response (Y)*, select the data from *Advertising Budget* and for *Predictor (X)* select *Sales*.

For the *Type of Regression* model select – *Linear*.

The regression analysis is then obtained with details on the regression equation, R-squared and the adjusted R-squared and analysis of variance (see Figure 6.10). The results obtained within this analysis are identical with those from the analysis carried our using Excel.

Figure 6.8 Excel data analysis – regression analysis

	A	B	C	D	E	F	G	H	I
1	SUMMARY OUTPUT								
2									
3	*Regression Statistics*								
4	Multiple R	0.986905279							
5	R Square	0.97398203							
6	Adjusted R Square	0.971091144							
7	Standard Error	10.92250277							
8	Observations	11							
9									
10	ANOVA								
11		*df*	*SS*	*MS*	*F*	*Significance F*			
12	Regression	1	40194.2904	40194.3	336.914761	1.93279E-08			
13	Residual	9	1073.709601	119.301					
14	Total	10	41268						
15									
16		*Coefficients*	*Standard Error*	*t Stat*	*P-value*	*Lower 95%*	*Upper 95%*	*Lower 95.0%*	*Upper 95.0%*
17	Intercept	6.122045131	6.642174473	0.92169	0.38073246	-8.90359743	21.1476877	-8.90359743	21.1476877
18	Sales	0.305764637	0.016658168	18.3552	1.9328E-08	0.268081243	0.34344803	0.26808124	0.34344803

Figure 6.9 Linear regression in Minitab – input data

	C1 Sales	C2 Advertising budget
1	87	30
2	150	45
3	164	50
4	210	65
5	245	88
6	298	95
7	365	120
8	379	150
9	589	187
10	635	192
11	687	210

Logarithmical, exponential and polynomial regression can be used in Excel, when data available for analysis displays a non-linear behaviour. Finding the regression function for data that is not linear, a similar approach can be followed. For example, for our data when an exponential function is selected the exponential equation is:

$$y = 33.499e^{0.003x} \text{ with the } R^2 = 0.8864$$

For a selection of the logarithmical function, we have the answer $y = 94.609ln(x) - 424.03$, with the $R^2 = 0.9385$. Where for data when a polynomial function is being selected, the equations is $y = -0.0002x^2 + 0.4615x - 16.5888$, with $R^2 = 0.9836$.

Multiple regression models

As with liner regression, multiple regression considers a dependant variable, and two or more independent variables. In many situations in a supply chain study, an analyst will be faced with working with more than one independent variable.

Figure 6.10 Linear regression in Minitab – regression analysis

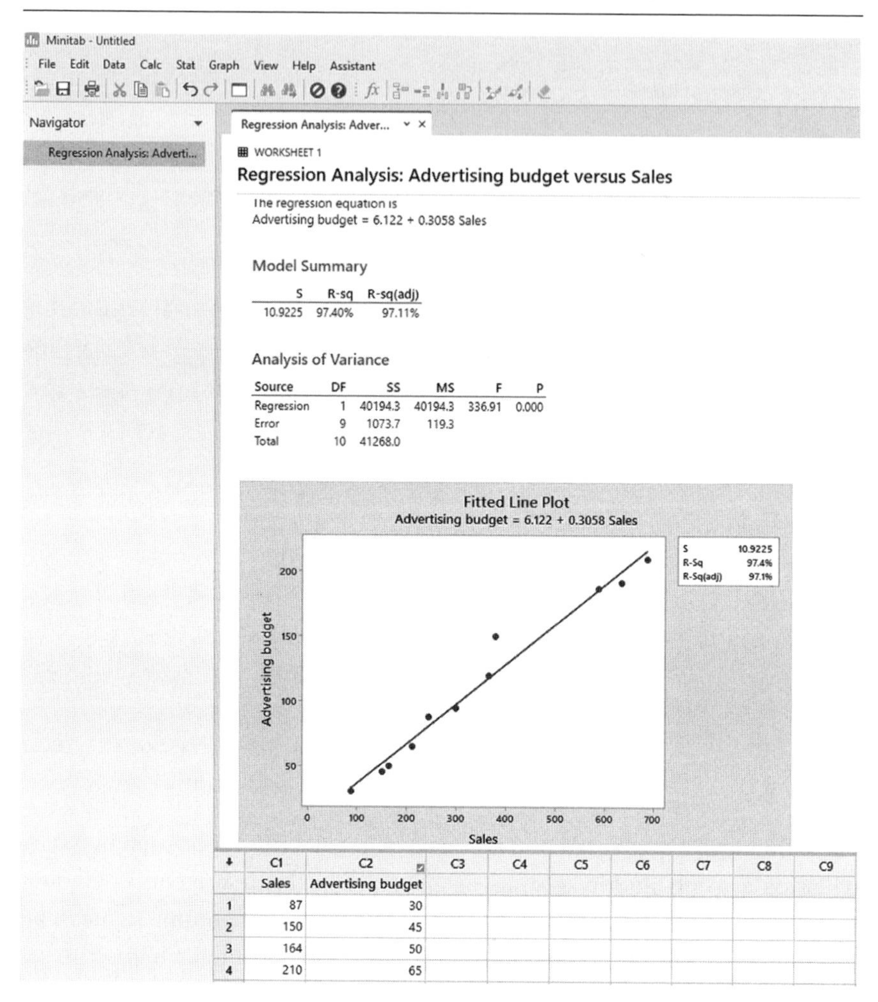

The multiple regression function can be extended from the linear function into

$$y = b_0 + b_1 * x_1 + b_2 * x_2 + \dots + b_n * x_n$$

In this particular equation, the y represents the dependent variable for which a value will be predicted, where the b_0, b_1, ... b_n, are the coefficients for which values will be identified based on the data analysed, or data available for the analysis.

The $x_1, x_2, \dots, x_n$ are used to indicate the independent variables. For these independent variables, numerical values are available or collected and used for the analysis, to identify a particular equation for y. When new values are being considered for the dependant variables, the identified equation is used to determine and to predict the new value for the independent variable y. The multiple regression model has been used in a number of research publications in the field of logistics and supply chain management, with authors working to identify, for example, different performance measures based on data already available to them (El-Sakty et al, 2014) or used in combination with other forecasting models such as seasonal autoregressive integrated moving average model in combination with multiple linear regression used in the food retail industry (Arunraj and Ahrens, 2015).

There are a number of other well-known regression algorithms available in the literature that can be applied to analysing data and big data. Details of these algorithms go beyond the purpose of this chapter; however, some of these are: logistics regression, ridge regression, stepwise regression and Lasso regression.

Forecasting: Time-series based models

A number of time-series based forecasting models can be categorized under this section with the most popular techniques being moving average (MA), single exponential smoothing (SES), double exponential smoothing (DES) and triple exponential smoothing (TES) all of which are detailed below. These techniques will use historical data collected over a period of time to be able to engage in prediction. However, there are a number of limitations when these forecasting techniques are to be employed. One key implication considers that the phenomena that have happened in the past will continue to happen in the future. One other factor considered in this case is that only time is involved in the decision process.

For any forecasting techniques considered for analysis, it is relevant to understand the forecasting accuracy. In this case, the following forecasting errors are to be considered:

Mean absolute deviation (MAD) with the formula set as:

$$MAD = \frac{1}{n}\sum_{t=1}^{n} |A_t - F_t|$$

Where:

n – is the number of observations used in the calculations

t – indicates the time now, so the current period of analysis

A_t – represents the numerical value for the actual data at current period of time t

F_t – is the forecast at period t

This formula will measure the absolute error.

Mean absolute percentage error (MAPE) will follow the following formula:

$$MAPE = \frac{100}{n} \sum_{t=1}^{n} \left| \frac{A_t - F_t}{A_t} \right|$$

and mean square error (MSE) is represented as:

$$MSE = \frac{1}{n} \sum_{t=1}^{n} (A_t - F_t)^2$$

Moving average

The moving average (MA) technique is a time-series forecasting technique that is easy to understand, and easy to implement with a mathematical model based on using averages to forecast. In other words, data collected from a number of past periods of time known as number of observations can be added together and divided by the number of observations.

$$MA_t = \frac{1}{n} \sum_{t=1}^{n} A_t$$

$$F_{t+1} = MA_t$$

Where:

n – is the number of observations used in the calculations of each average. Authors such as Albright and Winston (2011) have named this *span*. The selection of this value depends entirely on the analyst. However, if the selected value is '1', the results will simply indicate that the forecasted value is exactly the same as the current value. Larger values of n will indicate a smoother predicted value, where lower values will be able to capture some of the peaks and valleys in the data

t – indicates the time now, so the current period of analysis

$t+1$ – indicates time now plus one period ahead, therefore one period into the future

A_t – represents the numerical value for the actual data at current period of time t

F_t – the forecast at period t

F_{t+1} – the forecast for the time period $t+1$

To carry out a forecasting analysis using the moving average technique the following steps should be considered:

Step 1. Identify the need to forecast.

Step 2. Determine the time period for which to forecast. This step requires the analysis to indicate for how many periods ahead the forecast is planned to take place. For example, the requirement may be to forecast just one period ahead, or maybe two or three periods. As soon as the MA method is selected, the forecast ahead will be equal to the current value forecasted. Therefore, the results for 2–3 periods ahead will return a value equal to the last value calculated.

Step 3. Identify and collect the data required to carry out the forecast. Clean and evaluate the collected data. An analysis of the data collected is expected in this case regarding the number of observations, if all values over the observed period have been saved, if they are accurate, if there are any anomalies within the data collected.

Step 4. Evaluate whether moving average would be the most appropriate forecasting technique to be used for the selected data.

Step 5. Identify the value of '*n*' for which the MA should take place.

Step 6. If using Excel to forecast, develop the template that is the most appropriate for carrying out the forecast using MA (Figure 6.12).

Step 7. Calculate the MA and the forecast using the correct average formula. Calculate the forecasting errors (MAD, MAPE and MSE).

Step 8. Repeat steps 5 to 7 for different values of '*n*'.

Step 9. Construct the analysis table for different values of '*n*'. Plot the data and the forecast on a graph for each value of '*n*'.

Step 10. Analyse the results and provide a full report regarding the use of this technique as well as the final forecasted results.

To set a template (Step 6) for starting the analysis using the moving average technique, the minimum information required will be the time period used to collect the data, the actual data collected, a section to calculate the forecasted MA for a set period n, the forecast for this set period, the errors such as MAD, MAPE and MSE as depicted in Figure 6.12.

Considering we have actual demand data for a product such as Product 1 inserted in column D (see Figure 6.11), the forecasting analysis can start to take shape.

Figure 6.11 Product 1 data

Month	Product 1
Jan-17	1584
Feb-17	2676
Mar-17	1988
Apr-17	2202
May-17	2428
Jun-17	2012
Jul-17	2305
Aug-17	2060
Sep-17	2044
Oct-17	1241
Nov-17	2402
Dec-17	2346

Month	Product 1
Jan-18	2282
Feb-18	2686
Mar-18	2706
Apr-18	2196
May-18	2576
Jun-18	1972
Jul-18	2097
Aug-18	1953
Sep-18	2119
Oct-18	2085
Nov-18	2364
Dec-18	1648

Month	Product 1
Jan-19	1903
Feb-19	2055
Mar-19	2277
Apr-19	2191
May-19	1735
Jun-19	1969
Jul-19	2456
Aug-19	1842
Sep-19	2198
Oct-19	1274
Nov-19	1842
Dec-19	2611

Month	Product 1
Jan-20	1629
Feb-20	

In this particular example we have n = 3, which assumes that three previous observations are being used to carry out the calculations.

To carry out the analysis it is relevant to have the graphical representation of the data collected. The formula to be used in Excel in cell C4 to calculate MA3 (moving average for n=3) for the period March 2017 is = AVERAGE (B2:B4).

This formula can then be copied up to cell C38. To be able to identify the forecast using the formula

$$F_{t+1} = MA_t$$

the forecast for MA3 can be calculated in column D, where the value from cell D5 is made equal to the value calculated in cell C4. This formula can now to copied down to cell D39. Therefore, the forecast for February 2020 is now calculated as a value of 2027.33.

To support the analysis, the values required for the errors are essential in the analysis. The calculations for MAD(MA3) will start in the cell E5 and will use the Excel formula =ABS(B5-D5). Similarly, the calculations for MAPE(MA3) will start in cell F5 with the formula =ABS((B5-D5)/B5) and the MSE(MA3) will be in cell G5 with the formula =(B5-D5)^2.

All these formulas can now be copied up to cells E38, F38 and G38 respectively. Within cells E39, F39 and G39 the values can be averaged to give the final mean value for the forecast errors. The value in cell F39 is to be presented in percentages.

The moving average analysis for $n=3$ can be represented as shown in Figure 6.13.

Continuing the analysis in this way, further values for n can be given, therefore new forecasting values can be calculated for the same set of data (in this case demand data for Product 1).

The analysis for $n=5$ follows with the same set of formulas inserted in the Excel spreadsheet as depicted in Figure 6.14. An analysis table can now

Figure 6.12 Moving average template

	A	B	C	D	E	F	G
1	**Month**	**Product 1**	**MA3**	**Forecast(MA3)**	**MAD(MA3)**	**MAPE(MA3)**	**MSE(MA3)**
2	Jan-17	1584					
3	Feb-17	2676					
4	Mar-17	1988	=AVERAGE(B2:B4)				
5	Apr-17	2202	AVERAGE(number1, [number2], ...)				
6	May-17	2428					
7	Jun-17	2012					
8	Jul-17	2305					
9	Aug-17	2060					
10	Sep-17	2044					
11	Oct-17	1241					
12	Nov-17	2402					
13	Dec-17	2346					
14	Jan-18	2282					
15	Feb-18	2686					
16	Mar-18	2706					
17	Apr-18	2196					
18	May-18	2576					
19	Jun-18	1972					
20	Jul-18	2097					
21	Aug-18	1953					

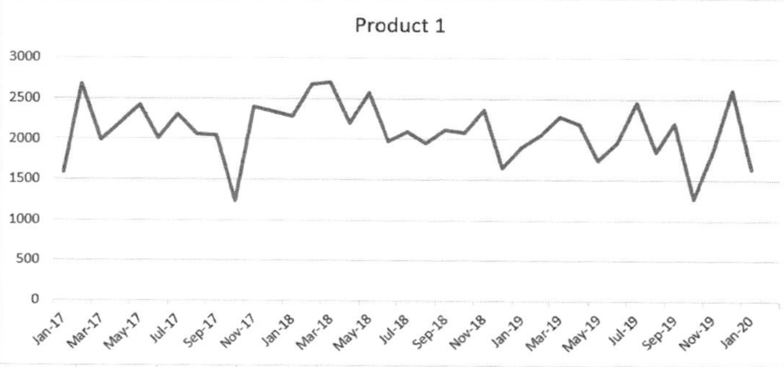

Product 1
3000
2500
2000
1500
1000
500
0
Jan-17 Mar-17 May-17 Jul-17 Sep-17 Nov-17 Jan-18 Mar-18 May-18 Jul-18 Sep-18 Nov-18 Jan-19 Mar-19 May-19 Jul-19 Sep-19 Nov-19 Jan-20

Figure 6.13 Moving average MA3

	A	B	C	D	E	F	G
1	Month	Product 1	MA3	Forecast(MA3)	MAD(MA3)	MAPE(MA3)	MSE(MA3)
2	Jan-17	1584					
3	Feb-17	2676					
4	Mar-17	1988	2082.67				
5	Apr-17	2202	2288.67	2082.67	119.33	0.05	14240.44
6	May-17	2428	2206.00	2288.67	139.33	0.06	19413.78
7	Jun-17	2012	2214.00	2206.00	194.00	0.10	37636.00
8	Jul-17	2305	2248.33	2214.00	91.00	0.04	8281.00
9	Aug-17	2060	2125.67	2248.33	188.33	0.09	35469.44
10	Sep-17	2044	2136.33	2125.67	81.67	0.04	6669.44
11	Oct-17	1241	1781.67	2136.33	895.33	0.72	801621.78
12	Nov-17	2402	1895.67	1781.67	620.33	0.26	384813.44
13	Dec-17	2346	1996.33	1895.67	450.33	0.19	202800.11
14	Jan-18	2282	2343.33	1996.33	285.67	0.13	81605.44
15	Feb-18	2686	2438.00	2343.33	342.67	0.13	117420.44
16	Mar-18	2706	2558.00	2438.00	268.00	0.10	71824.00
17	Apr-18	2196	2529.33	2558.00	362.00	0.16	131044.00
18	May-18	2576	2492.67	2529.33	46.67	0.02	2177.78
19	Jun-18	1972	2248.00	2492.67	520.67	0.26	271093.78
20	Jul-18	2097	2215.00	2248.00	151.00	0.07	22801.00
21	Aug-18	1953	2007.33	2215.00	262.00	0.13	68644.00
22	Sep-18	2119	2056.33	2007.33	111.67	0.05	12469.44
23	Oct-18	2085	2052.33	2056.33	28.67	0.01	821.78
24	Nov-18	2364	2189.33	2052.33	311.67	0.13	97136.11
25	Dec-18	1648	2032.33	2189.33	541.33	0.33	293041.78
26	Jan-19	1903	1971.67	2032.33	129.33	0.07	16727.11
27	Feb-19	2055	1868.67	1971.67	83.33	0.04	6944.44
28	Mar-19	2277	2078.33	1868.67	408.33	0.18	166736.11
29	Apr-19	2191	2174.33	2078.33	112.67	0.05	12693.78
30	May-19	1735	2067.67	2174.33	439.33	0.25	193013.78
31	Jun-19	1969	1965.00	2067.67	98.67	0.05	9735.11
32	Jul-19	2456	2053.33	1965.00	491.00	0.20	241081.00
33	Aug-19	1842	2089.00	2053.33	211.33	0.11	44661.78
34	Sep-19	2198	2165.33	2089.00	109.00	0.05	11881.00
35	Oct-19	1274	1771.33	2165.33	891.33	0.70	794475.11
36	Nov-19	1842	1771.33	1771.33	70.67	0.04	4993.78
37	Dec-19	2611	1909.00	1771.33	839.67	0.32	705040.11
38	Jan-20	1629	2027.33	1909.00	280.00	0.17	78400.00
39	Feb-20			**2027.33**	**299.30**	**15.65%**	**146100.24**
40							

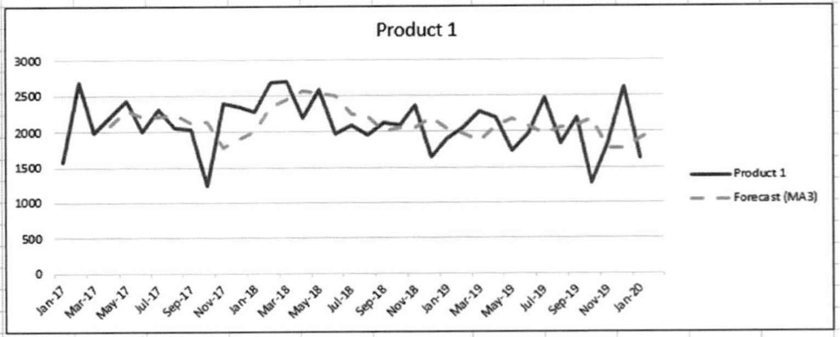

Analysis table

n	Forecast	MAD	MAPE	MSE
3	2027.33	299.30	15.65%	146100.24

Figure 6.14 Moving average MA3 and MA5

	A	B	C	D	E	F	G	I	J	K	L	M
1	Month	Product 1	MA3	Forecast(MA3)	MAD(MA3)	MAPE(MA3)	MSE(MA3)	MA5	Forecast(MA5)	MAD(MA5)	MAPE(MA5)	MSE(MA5)
2	Jan-17	1584										
3	Feb-17	2676										
4	Mar-17	1988	2082.67									
5	Apr-17	2202	2288.67	2082.67	119.33	0.05	14240.44					
6	May-17	2428	2206.00	2288.67	139.33	0.06	19413.78	2175.6				
7	Jun-17	2012	2214.00	2206.00	194.00	0.10	37636.00	2261.2	2175.6	163.6	0.081	26764.96
8	Jul-17	2305	2248.33	2214.00	91.00	0.04	8281.00	2187	2261.2	43.8	0.019	1918.44
9	Aug-17	2060	2125.67	2248.33	188.33	0.09	35469.44	2201.4	2187	127	0.062	16129
10	Sep-17	2044	2136.33	2125.67	81.67	0.04	6669.44	2169.8	2201.4	157.4	0.077	24774.76
11	Oct-17	1241	1781.67	2136.33	895.33	0.72	801621.78	1932.4	2169.8	928.8	0.748	862669.44
12	Nov-17	2402	1895.67	1781.67	620.33	0.26	384813.44	2010.4	1932.4	469.6	0.196	220524.16
13	Dec-17	2346	1996.33	1895.67	450.33	0.19	202800.11	2018.6	2010.4	335.6	0.143	112627.36
14	Jan-18	2282	2343.33	1996.33	285.67	0.13	81605.44	2063	2018.6	263.4	0.115	69379.56
15	Feb-18	2686	2438.00	2343.33	342.67	0.13	117420.44	2191.4	2063	623	0.232	388129
16	Mar-18	2706	2558.00	2438.00	268.00	0.10	71824.00	2484.4	2191.4	514.6	0.190	264813.16
17	Apr-18	2196	2529.33	2558.00	362.00	0.16	131044.00	2443.2	2484.4	288.4	0.131	83174.56
18	May-18	2576	2492.67	2529.33	46.67	0.02	2177.78	2489.2	2443.2	132.8	0.052	17635.84
19	Jun-18	1972	2248.00	2492.67	520.67	0.26	271093.78	2427.2	2489.2	517.2	0.262	267495.84
20	Jul-18	2097	2215.00	2248.00	151.00	0.07	22801.00	2309.4	2427.2	330.2	0.157	109032.04
21	Aug-18	1953	2007.33	2215.00	262.00	0.13	68644.00	2158.8	2309.4	356.4	0.182	127020.96
22	Sep-18	2119	2056.33	2007.33	111.67	0.05	12469.44	2143.4	2158.8	39.8	0.019	1584.04
23	Oct-18	2085	2052.33	2056.33	28.67	0.01	821.78	2045.2	2143.4	58.4	0.028	3410.56
24	Nov-18	2364	2189.33	2052.33	311.67	0.13	97136.11	2123.6	2045.2	318.8	0.135	101633.44
25	Dec-18	1648	2032.33	2189.33	541.33	0.33	293041.78	2033.8	2123.6	475.6	0.289	226195.36
26	Jan-19	1903	1971.67	2032.33	129.33	0.07	16727.11	2023.8	2033.8	130.8	0.069	17108.64
27	Feb-19	2055	1868.67	1971.67	83.33	0.04	6944.44	2011	2023.8	31.2	0.015	973.44
28	Mar-19	2277	2078.33	1868.67	408.33	0.18	166736.11	2049.4	2011	266	0.117	70756
29	Apr-19	2191	2174.33	2078.33	112.67	0.05	12693.78	2014.8	2049.4	141.6	0.065	20050.56
30	May-19	1735	2067.67	2174.33	439.33	0.25	193013.78	2032.2	2014.8	279.8	0.161	78288.04
31	Jun-19	1969	1965.00	2067.67	98.67	0.05	9735.11	2045.4	2032.2	63.2	0.032	3994.24
32	Jul-19	2456	2053.33	1965.00	491.00	0.20	241081.00	2125.6	2045.4	410.6	0.167	168592.36
33	Aug-19	1842	2089.00	2053.33	211.33	0.11	44661.78	2038.6	2125.6	283.6	0.154	80428.96
34	Sep-19	2198	2165.33	2089.00	109.00	0.05	11881.00	2040	2038.6	159.4	0.073	25408.36
35	Oct-19	1274	1771.33	2165.33	891.33	0.70	794475.11	1947.8	2040	766	0.601	586756
36	Nov-19	1842	1771.33	1771.33	70.67	0.04	4993.78	1922.4	1947.8	105.8	0.057	11193.64
37	Dec-19	2611	1909.00	1771.33	839.67	0.32	705040.11	1953.4	1922.4	688.6	0.264	474169.96
38	Jan-20	1629	2027.33	1909.00	280.00	0.17	78400.00	1910.8	1953.4	324.4	0.199	105235.36
39	Feb-20			**2027.33**	**299.30**	**15.65%**	**146100.24**		**1910.8**	**306.11**	**15.92%**	**142745.88**

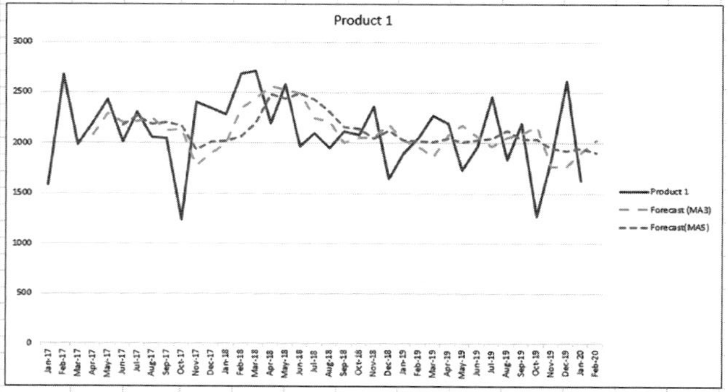

Analysis table

n	Forecast	MAD	MAPE	MSE
3	2027.33	299.30	15.65%	146100
5	1910.8	306.11	15.92%	142746

be constructed to capture the information regarding the forecasted value as well as the errors, such as MAD (MA5), MAPE (MA5) and MSE (MA5).

Following these steps, the analysis can continue for more values of n. However, in many practical cases analysts may decide to restrict the MA calculations to just two observations.

Moving average analysis using Minitab 19

Sales data for Product 1 is inserted in the spreadsheet section of the Minitab software (see Figure 6.15). Using a scatterplot, the data can be displayed in a graph as follows:

> *Graph > Scatterplot > With Connect Line*, where for the Y variables select Product 1 and for the X variable select Month.

Figure 6.15 Product 1 data in Minitab

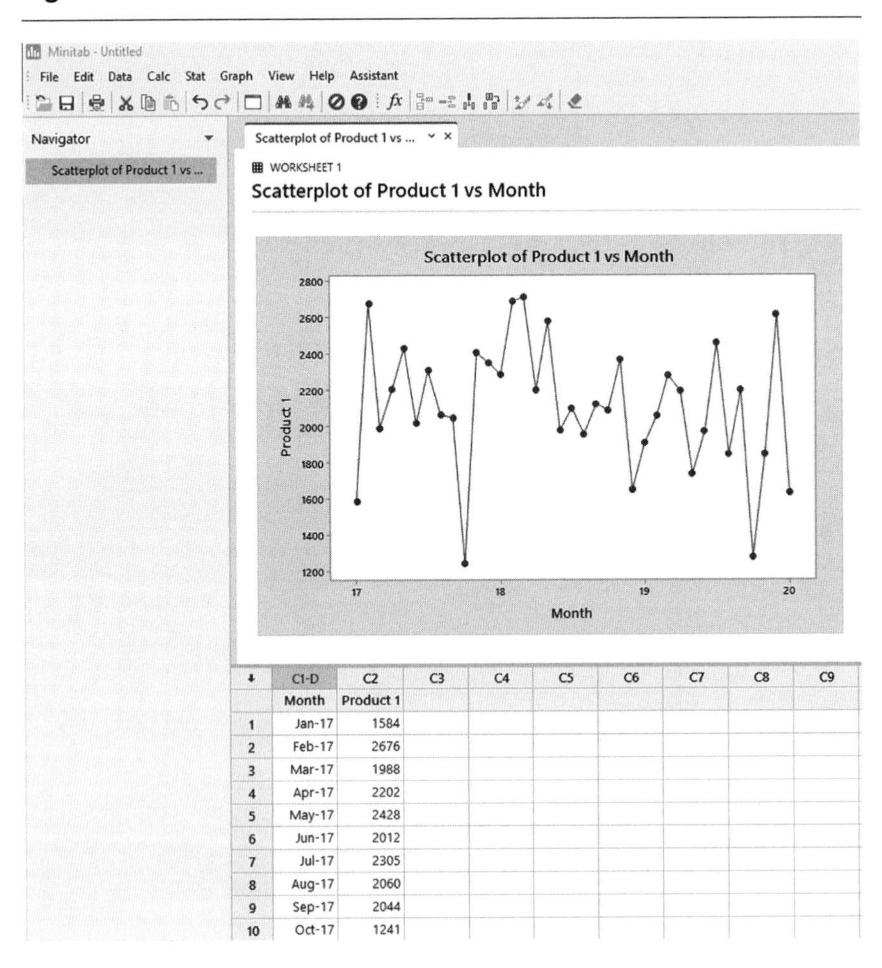

To carry out the analysis for the moving average, the following instructions can be followed:

Stat > Time Series > Moving Average

A new screen appears, where for the section variable, select Product 1 and for the section MA length select Value 3.

Select 'Generate Forecast' with the number of forecasts as 1 followed by 'OK'. The analysis for MA3 is now provided with the values as indicated in Figure 6.16.

Figure 6.16 Product 1 MA3 in Minitab

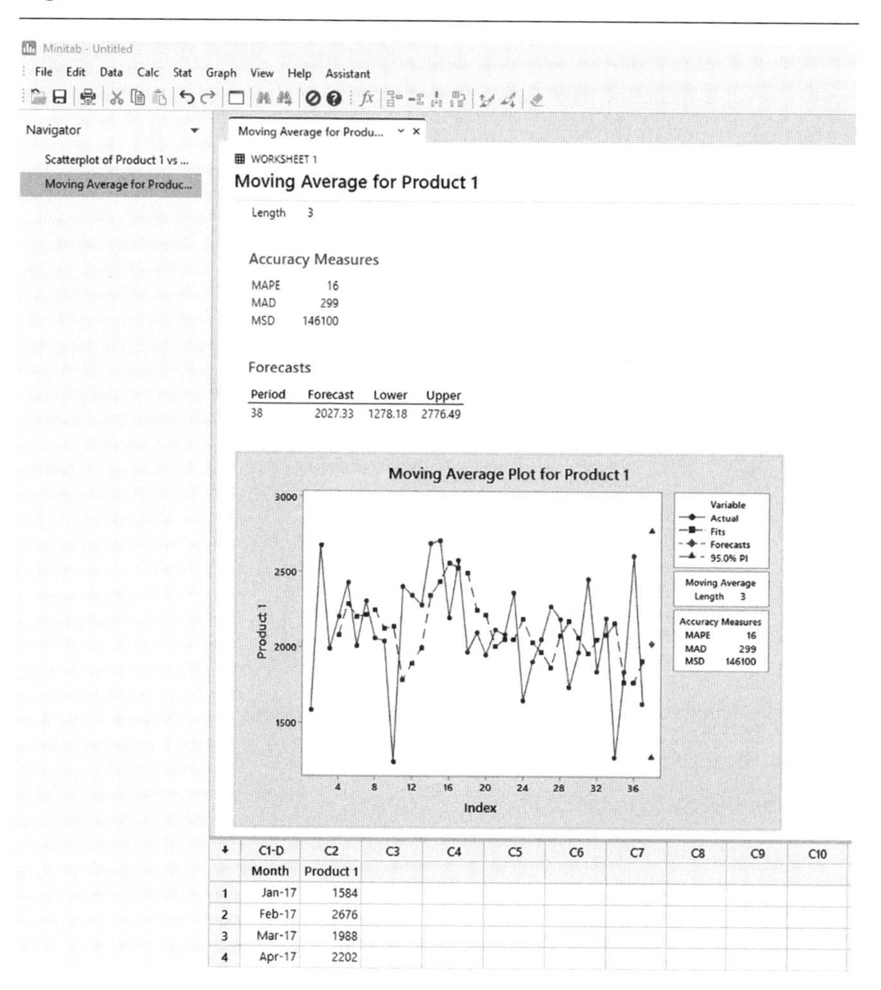

Similarly, this analysis can be carried out for MA5 and any other values for n.

Within Minitab 19 there is also the option to 'Center the Moving Averages'. If this is selected, a change can be observed in the graphical representation and the final results as indicated in Figure 6.17. The results obtained for MAPE MAD and MSD are all lower than in the previous case.

Some of the disadvantages linked with the MA technique are related to the data behaviour, in other words with the shape of the data collected for analysis. If there is clear 'noise' in data, variations in data, MA does not cope well as it tends to smooth the data out. MA is more reliable for short-term forecasts.

Single exponential smoothing

Single exponential smoothing (SES) is another forecasting technique that puts emphasis on weights of past data (Albright and Winston, 2011). However, it is an effective and at the same time versatile technique (Evans, 2016).

To introduce the forecasting model for this particular technique, the term *level* (L) is being introduced. This term relates to the level where data would be if there were no random noise in the data (Albright and Winston, 2011). One other term that appears in the definition of this technique is the smoothing coefficient alpha (α). This coefficient can have a value between 0 and 1. The model presented below has been used for a number of years and it is similarly presented in Albright and Winston (2011, p 884).

$$L_t = \alpha * A_t + (1 - \alpha) * L_{t-1}$$

$$F_{t+1} = L_t$$

where:

A_t – represents the numerical value for the actual data at current period of time t

L_t – is the level value at current time

L_{t-1} – is the level value at previous time period

F_t – is the forecast at period t

F_{t+1} – is the value forecasted at period $t+1$

α – is the smoothing coefficient, which ranges from 0 to 1

t – indicates the time now, so the current period of analysis

$t+1$ – indicates time now plus one period ahead, therefore one period into the future

Figure 6.17 Product 1 centred MA3 in Minitab

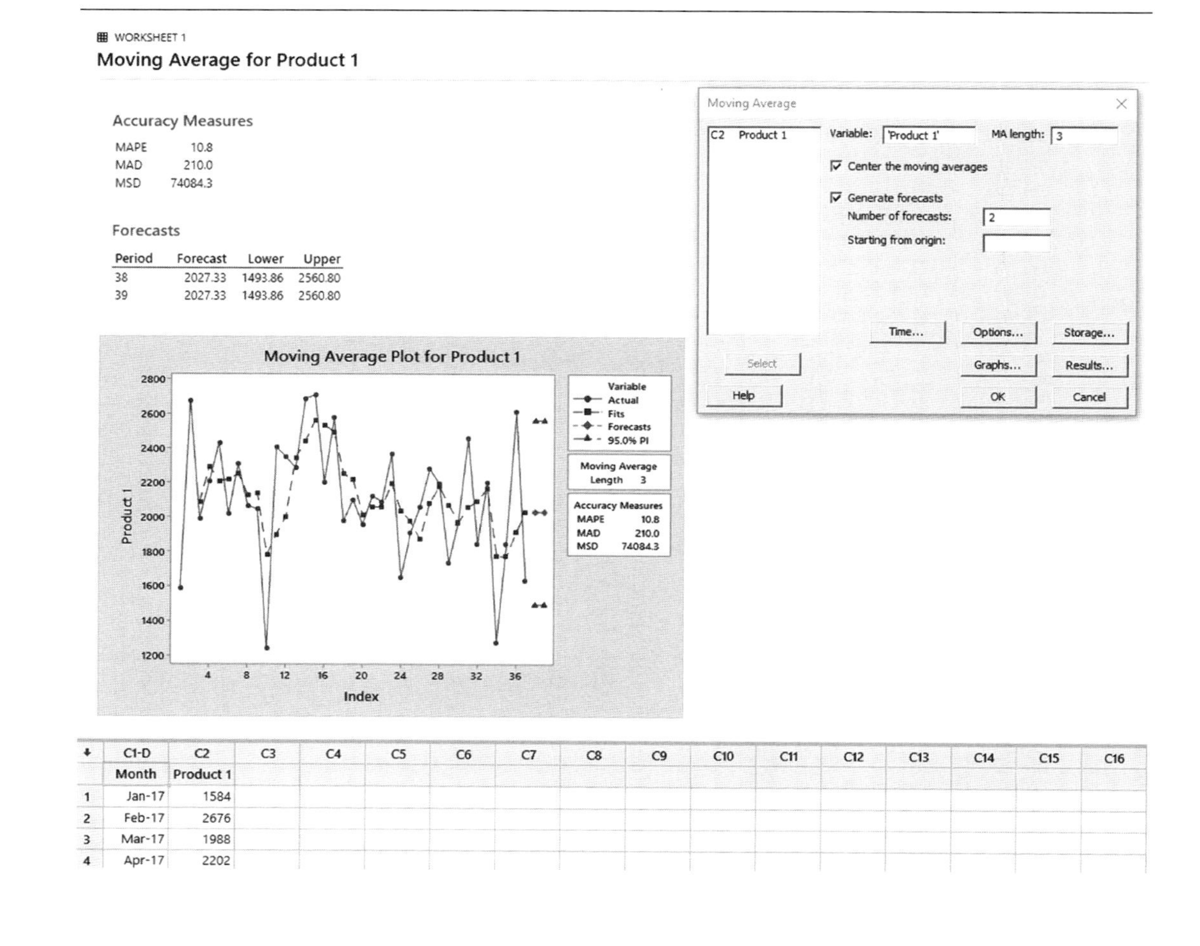

The formula presented above can be expressed as:

*Current level = alpha * Actual data + (1- alpha) * Previous level*
Forecast at (t+1) = Current level

To carry out a forecasting analysis using the SES technique the following steps should be considered:

Step 1. Identify the need to forecast.

Step 2. Determine the time period for which to forecast. This step requires the analysis to indicate for how many periods ahead the forecast is planned to take place.

Step 3. Identify and collect the data required to forecast. Clean and evaluate the collected data. An analysis of the collected data is expected in this case regarding the number of observations, if all values over the observed period have been saved, if they are accurate, if there are any anomalies within the data collected.

Step 4. Evaluate whether SES would be the most appropriate forecasting technique to be used for the selected data.

Step 5. Identify the value of alpha coefficient α for which the SES should take place. Values for $\alpha \in (0, 1)$.

Step 6. Identify the L_0 by using one of these three methods:

a. $L_0 = A_1$ (level at time 0 is the same as the first selected actual data)

b. L_0 = the corresponding value from the trend equation

c. L_0 = with the average of the first 'few' values

This selection depends on the data available and it is the decision of the analyst working with this data.

Step 7. In Excel, develop the template required for SES. Calculate the *level* for the SES and the forecasting using the correct formulas. Calculate the forecasting errors (MAD, MAPE and MSE).

Step 8. Repeat Steps 5 to 7 for different values of 'α'.

Step 9. Construct the analysis table for different values of 'α' and L_0. Plot the data and the forecast on a graph for each calculated value of 'α'.

Step 10. Analyse the results and provide a full report regarding the use of this technique as well as the final forecasted results.

Analysis Assuming Step 1 is being considered and sales data for Product 1 is put forward to forecast. Step 2, the time horizon decided in this case is two months. So, the analysis is to be carried out for a short time. Step 3, the data to be collected and forecasted is still the sales data for Product 1. Step 4, the MA analysis has been conducted for this product. However, now the SES is being considered for the same product. Step 5, the decision is to continue with the single exponential smoothing method and therefore a value of α is set to be 0.3. To start the analysis, this value can be selected as random; however, this will need to be adjusted as the analysis is being completed.

Step 6. To calculate the value of L_0, so the starting point for the level for option a the analyst can select the value 1584. For this particular set of data, this selection may not be the best as this value is not between the best averages for the initial set of data. If the decision is to go with option b in this case the trend equation identified for this set of data is '$= -6.8385x + 2236.8$', therefore the value of 2236.8 can be selected as a starting point for L_0. However, if the selection is to go with option c the value of L_0 can be of the first six months, therefore the value of 2148.33. In this particular case, the value selected to demonstrate this example is option c. However, an analyst should try every option to identify which will generate the smallest value of errors.

Step 7. In Excel a template as indicated in Figure 6.18 can be developed. The value of alpha will be inserted in cell C1, where the value of L_0 is to be inserted in cell C3. The equation for L_1 is inserted in cell C4 following the formula from equation... where in Excel it will be represented as =$C$1*B4+(1-$C$1)*C3. Cell C1 is locked as the value for alpha remains the same for this particular set of calculations.

Figure 6.18 Single exponential smoothing template setting

	A	B	C	D	E	F	G
1			0.3				
2	Month	Product 1	Level	Forecast	MAD(SES)	MAPE(SES)	MSE(SES)
3			2148.33				
4	Jan-17	1584	=$C$1*B4+(1-$C$1)*C3				
5	Feb-17	2676					
6	Mar-17	1988					
7	Apr-17	2202					
8	May-17	2428					
9	Jun-17	2012					
10	Jul-17	2305					
11	Aug-17	2060					

Continuing with the development of this template, in cell D4 the following equations is inserted '=C3' following the formula provided earlier. Cell E4 has the formula: =ABS(B4-D4); cell F4 is populated with =ABS((B4-D4)/B4) and cell G4 with =(B4-D4)^2.

These functions can now be copied as follows. The function inserted in C4, E4, F4 and G4 can be copied up to cells C40, E40, F40 and G40. However, the formula inserted in D4 can be copied up to cell D41. To complete the calculations for MAD, MAPE and MSE, in cells E41, F41 and G41 the average functions are used such as: E41=AVERAGE(E4:E40); F41=AVERAGE(F4:F40) and G41 =AVERAGE(G4:G40). These will indicate the averages, and therefore the final values for the set of data for Mad, MAPE and MSE.

The complete analysis for $\alpha = 0.3$ and $L_0 = 2148.33$ is shown in Figure 6.19. Still, note that the analysis is not yet complete. To complete this analysis a range of values for alpha and L_0 are to be considered. These should be all captured within an analysis table for different values of the alpha coefficients such as 0.5 and 0.6 (Step 8). The example considered for $\alpha = 0.3$; $\alpha = 0.5$ and $\alpha = 0.7$ is now presented in Figure 6.20 (Step 9).

Step 10. An analysis of the results takes place when a range of values has been calculated. In this particular case the lowest error values for all MAD, MAPE and MSE have been obtained in the case of $\alpha = 0.3$. Therefore, the forecasted value for February 2020 is confirmed as a value of 1941.07.

A similar analysis can be carried out using Minitab.

Single exponential smoothing analysis using Minitab 19

We still have sales data for Product 1 inserted in the spreadsheet section of the Minitab software as in Figure 6.15. The instructions to follow to carry out a single exponential smoothing analysis are:

Stat > Time Series > Single Exp Smoothing...

A new screen appears, where for the section *Variable*, select *Product 1* and for the section *Weight to Use in Smoothing* use *0.3*.

Select *Generate Forecast* with the number of forecasts as 2 followed by *OK*. The analysis for SES (0.3) is now provided with the values as indicated in Figure 6.21. The results obtained in this case are the same with those obtained in the Excel analysis. What is interesting to note in this case is the selection of the L_0. For any other selection, the final results would have had a slightly different answer when the results from the two selected software packages are compared.

Figure 6.19 Product 1 single exponential smoothing and alpha 0.3

Month	Product 1	Level	Forecast (0.3)	MAD(SES)	MAPE(SES)	MSE(SES)
		0.3				
		2148.33				
Jan-17	1584	1979.03	2148.33	564.33	0.36	318472.11
Feb-17	2676	2188.12	1979.03	696.97	0.26	485762.53
Mar-17	1988	2128.09	2188.12	200.12	0.10	40049.35
Apr-17	2202	2150.26	2128.09	73.91	0.03	5463.23
May-17	2428	2233.58	2150.26	277.74	0.11	77139.27
Jun-17	2012	2167.11	2233.58	221.58	0.11	49098.72
Jul-17	2305	2208.48	2167.11	137.89	0.06	19014.31
Aug-17	2060	2163.93	2208.48	148.48	0.07	22044.92
Sep-17	2044	2127.95	2163.93	119.93	0.06	14383.86
Oct-17	1241	1861.87	2127.95	886.95	0.71	786685.47
Nov-17	2402	2023.91	1861.87	540.13	0.22	291743.62
Dec-17	2346	2120.53	2023.91	322.09	0.14	103743.95
Jan-18	2282	2168.97	2120.53	161.47	0.07	26071.00
Feb-18	2686	2324.08	2168.97	517.03	0.19	267315.48
Mar-18	2706	2438.66	2324.08	381.92	0.14	145861.30
Apr-18	2196	2365.86	2438.66	242.66	0.11	58882.64
May-18	2576	2428.9	2365.86	210.14	0.08	44158.73
Jun-18	1972	2291.83	2428.90	456.90	0.23	208759.58
Jul-18	2097	2233.38	2291.83	194.83	0.09	37959.32
Aug-18	1953	2149.27	2233.38	280.38	0.14	78614.10
Sep-18	2119	2140.19	2149.27	30.27	0.01	916.12
Oct-18	2085	2123.63	2140.19	55.19	0.03	3045.63
Nov-18	2364	2195.74	2123.63	240.37	0.10	57777.23
Dec-18	1648	2031.42	2195.74	547.74	0.33	300021.00
Jan-19	1903	1992.89	2031.42	128.42	0.07	16491.49
Feb-19	2055	2011.53	1992.89	62.11	0.03	3857.22
Mar-19	2277	2091.17	2011.53	265.47	0.12	70476.76
Apr-19	2191	2121.12	2091.17	99.83	0.05	9966.47
May-19	1735	2005.28	2121.12	386.12	0.22	149086.69
Jun-19	1969	1994.4	2005.28	36.28	0.02	1316.40
Jul-19	2456	2132.88	1994.40	461.60	0.19	213076.82
Aug-19	1842	2045.61	2132.88	290.88	0.16	84610.18
Sep-19	2198	2091.33	2045.61	152.39	0.07	23221.25
Oct-19	1274	1846.13	2091.33	817.33	0.64	668028.92
Nov-19	1842	1844.89	1846.13	4.13	0.00	17.07
Dec-19	2611	2074.72	1844.89	766.11	0.29	586921.66
Jan-20	1629	1941.01	2074.72	445.72	0.27	198670.16
Feb-20			1941.01	308.80	15.97%	147803.37

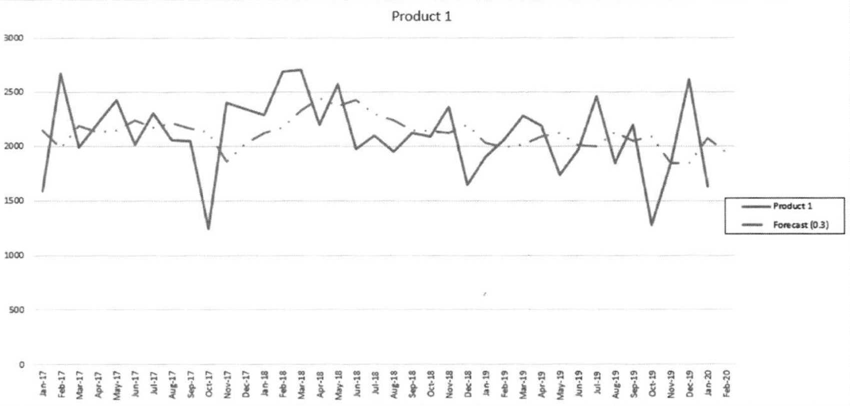

Product 1

Product 1
Forecast (0.3)

Figure 6.20 Product 1 single exponential smoothing analysis

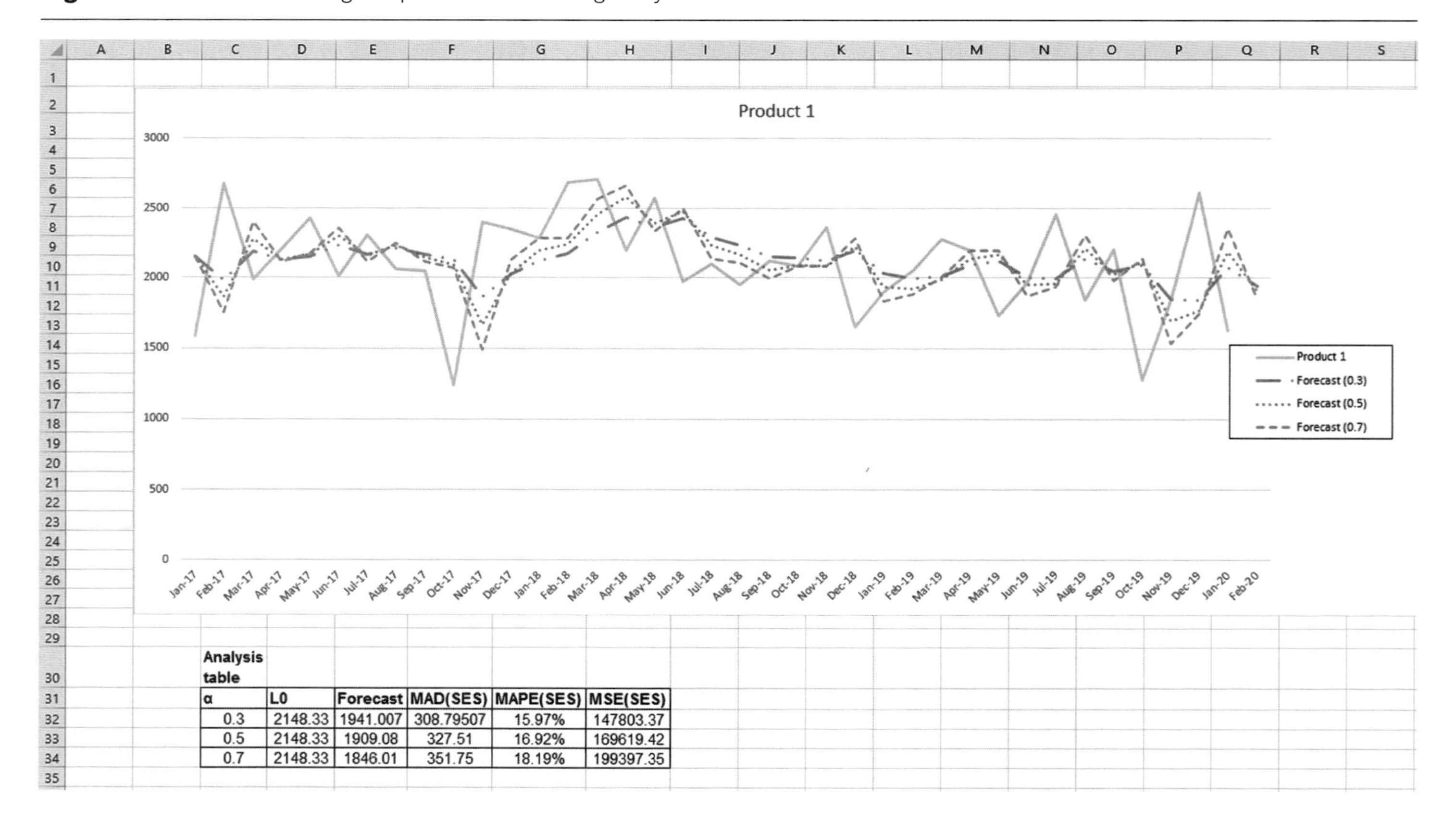

Analysis table

α	L0	Forecast	MAD(SES)	MAPE(SES)	MSE(SES)
0.3	2148.33	1941.007	308.79507	15.97%	147803.37
0.5	2148.33	1909.08	327.51	16.92%	169619.42
0.7	2148.33	1846.01	351.75	18.19%	199397.35

Figure 6.21 Product 1 single exponential smoothing in Minitab for alpha 0.3

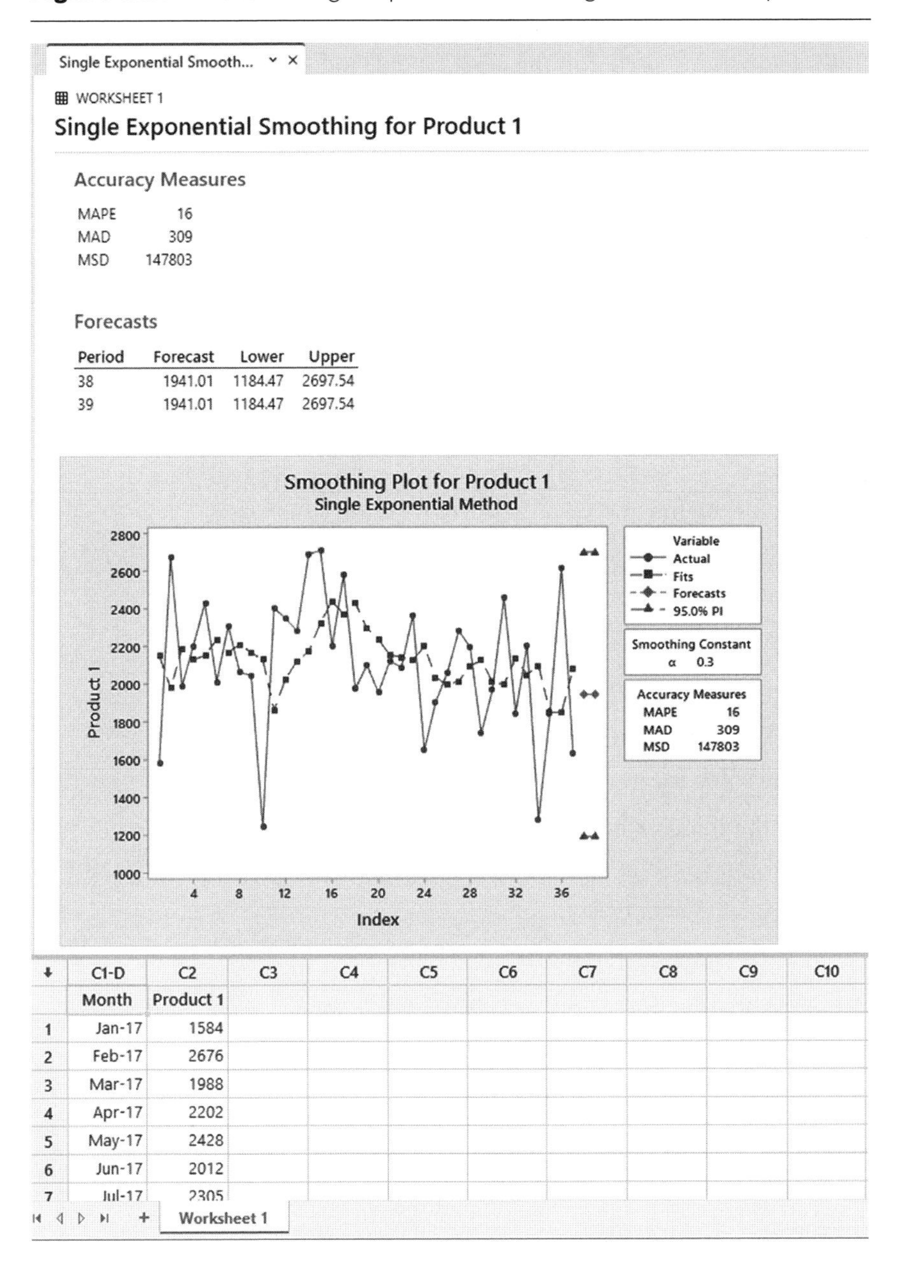

Double exponential smoothing

Double exponential smoothing (DES) is used for data that has trend, but not seasonality. This method is also called Holt's method.

The forecasting function represented in this technique consists of a level, trend and two smoothing coefficients.

The equations used for the level, trend and forecasting are similarly presented in Albright and Winston (2011, p 888) as below:

$$L_t = \alpha * A_t + (1 - \alpha) * (L_{t-1} + T_{t-1})$$
$$T_t = \beta * (L_t - L_{t-1}) + (1 - \beta) * T_{t-1}$$
$$F_t = L_{t-1} + T_{t-1}$$
$$F_{t+k} = L_t + k * T_t$$

where:

A_t – represents the numerical value for the actual data at current period of time t

t – indicates the time now, so the current period of analysis

$t+1$– indicates time now plus one period ahead, therefore one period into the future

$t+k$ – represents k periods into the future, where $k = 1, 2, 3, \ldots$

L_t – is the level value at current time

L_{t-1} – is the level value at previous time period

T_t – is the value for trend at time now

T_{t-1} – is the trend value at the previous period of time

F_t – is the forecast at period t. This is being calculated for all the values where actual data exists

F_{t+1} – is the value forecasted at period $t+1$

F_{t+k} – is the value forecasted at period $t+k$,

for $k = 1$, $F_{t+1} = L_t + 1 * T_t$

and for $k=2$, $F_{t+2} = L_t + 2 * T_t$

and so on.

α – is the smoothing coefficient corresponding with the level, which ranges from 0 to 1

β – is the smoothing coefficient corresponding with the trend, which ranges from 0 to 1

To carry out a forecasting analysis using the double exponential smoothing method the following stages can be considered:

Step 1. Identify the need to forecast.

Step 2. Determine the time period for which to forecast. This step requires the analysis to indicate for how many periods ahead the forecast is planned to take place.

Step 3. Identify and collect the data required to forecast. Clean and evaluate the collected data. An analysis of the data collected is expected in this case regarding the number of observations. This considers if all values over the observed period have been saved, if they are accurate, if there are any anomalies within the data collected.

Step 4. Evaluate whether double exponential smoothing would be the most appropriate forecasting technique to be used for the selected data.

Step 5. Select the most appropriate value of alpha α and beta β coefficients for which the DES should take place. Values for $\alpha, \beta \varepsilon (0, 1)$. For example, select the value for α and β between 0.1, 0.3, 0.5, 0.7 and 0.9. Note the values selected for α and β will not necessarily have the same value.

Step 6. Identify the L_0 by using one of the following three methods:

a. $L_0 =$ A1 (level at time 0 is the same as the first selected actual data)

b. $L_0 =$ the corresponding value from the trend equation

c. $L_0 =$ with the average of the first 'few' values

For T_0 use the corresponding value from the trend equation.

Step 7. In Excel, develop the template required for DES. Calculate the *level* and *trend* for the DES and the forecasting using the correct formulas. Calculate the forecasting errors (MAD, MAPE and MSE).

Step 8. Repeat Steps 5 to 7 for different values of 'α' and 'β'.

Step 9. Construct the analysis table for different values of alpha α and beta β and L_0 and T_0. Plot the data and the forecast on a graph for each calculated value of alpha α and beta β.

Step 10. Analyse the results and provide a full report regarding the use of this technique as well as the final forecasted results.

Analysis It has been identified that a forecasting analysis is required for Product 2 for the following week (Step 1 and Step 2). The sales data for 51 weeks has been identified and collected (Step 3). Trend in data has been observed, therefore double exponential smoothing (DES) has been selected for the analysis (Step 4).

Step 5. For the data collected, to start the analysis, a random selection for α and β is possible. This is with the mention that these values are to be changed later on during this analysis. Therefore, the starting step could be for $\alpha = 0.5$ and $\beta = 0.3$.

Step 6. For this particular case, we can use the trend equation with the value for $L_0 = 103.71$ and $T_0 = 5.89$.

Step 7. To start the analysis a DES template has been developed, as indicated in Figure 6.22. In cell C4 insert the function = $C$1*B4 + (1-$C$1)*(C3+D3) that represents the level equation. In cell D4 insert = $D$1*(C4-C3) + (1-$D$1)*D3 that represents the trend equations for the DES model.

The forecasting formula to be used is in cell E4 = C3+D3. The errors will also be inserted such as MAD in cell F4 = ABS(B4-E4); MAPE = ABS((B4-E4)/B4); and MSE = (B4-E4)^2.

Steps 8 and 9. Different values for α and β can be selected and they can be represented in the analysis table, as depicted in Figure 6.23.

Step 10. It can be noted here that for $\alpha = 0.5$ and $\beta = 0.3$ with a forecasted value of 468.14, the MSE gives the lowest value. However, for the MAD and MAPE there is a lower value in the case of $\alpha = 0.7$ and $\beta = 0.5$ with the forecasted value of 480.39. The difference in this case is marginally lower. However, this can make a big impact on the selected forecasted final value.

Double exponential smoothing analysis using Minitab 19

For the sales data for Product 2 inserted in the spreadsheet section of the Minitab software as in Figure 6.24. The instructions to follow to carry out double exponential smoothing analysis are:

Stat > Time Series > Double Exp Smoothing...

A new screen appears, where for the section *Variable*, select *Product 2* and for the section *Specified Weights* use 0.5 for level and 0.3 for trend.

Select *Generate Forecast* with the number of forecasts as 1 followed by *OK*. The analysis for DES (0.5 and 0.3) is now provided with the values as indicated in Figure 6.24. The results obtained in this case are the same with those obtained in the Excel analysis.

Figure 6.22 Product 2 double exponential smoothing template

	A	B	C	D	E	F	G	H
1			0.5	0.3				
2	Weeks	Product 2	Level	Trend	Forecast (	MAD(DES)	MAPE(DES)	MSE(DES)
3			103.71	5.89				
4	1	138						
5	2	147						
6	3	150						
7	4	149						
8	5	156						
9	6	153						
10	7	155						
11	8	163						
12	9	168						
13	10	165						
14	11	172						
15	12	169						
16	13	179						
17	14	182						
18	15	180						
19	16	185						
20	17	214						
21	18	189						
22	19	196						
23	20	200						

Figure 6.23 Product 2 DES with analysis table

	A	B	C	D	E	F	G	H
1			0.5	0.3				
2	Weeks	Product 2	Level	Trend	Forecast (	MAD(DES)	MAPE(DES)	MSE(DES)
3			103.64	5.89				
4	1	138	123.77	10.16	109.53	28.47	0.21	810.54
5	2	147	140.46	12.12	133.93	13.07	0.09	170.94
6	3	150	151.29	11.73	152.58	2.58	0.02	6.68
7	4	149	156.01	9.63	163.03	14.03	0.09	196.73
8	5	156	160.82	8.18	165.64	9.64	0.06	92.99
9	6	153	161.00	5.78	169.01	16.01	0.10	256.17
10	7	155	160.89	4.02	166.79	11.79	0.08	138.90
11	8	163	163.95	3.73	164.91	1.91	0.01	3.64
12	9	168	167.84	3.78	167.68	0.32	0.00	0.10
13	10	165	168.31	2.78	171.62	6.62	0.04	43.80
14	11	172	171.55	2.92	171.09	0.91	0.01	0.82
15	12	169	171.73	2.10	174.47	5.47	0.03	29.88
16	13	179	176.42	2.87	173.83	5.17	0.03	26.70
17	14	182	180.65	3.28	179.29	2.71	0.01	7.34
18	15	180	181.96	2.69	183.93	3.93	0.02	15.42
19	16	185	184.83	2.74	184.66	0.34	0.00	0.12
20	17	214	200.79	6.71	187.57	26.43	0.12	698.45
21	18	189	198.25	3.93	207.49	18.49	0.10	342.03
22	19	196	199.09	3.01	202.18	6.18	0.03	38.20
23	20	200	201.05	2.69	202.10	2.10	0.01	4.40
24	21	200	201.87	2.13	203.74	3.74	0.02	13.99
25	22	205	204.50	2.28	204.00	1.00	0.00	1.00
26	23	208	207.39	2.46	206.78	1.22	0.01	1.48
27	24	210	209.93	2.49	209.85	0.15	0.00	0.02
28	25	243	227.71	7.07	212.41	30.59	0.13	935.58
29	26	252	243.39	9.66	234.78	17.22	0.07	296.53

Analysis table

Alpha	Beta	Forecast	MAD(DES)	MAPE(DES)	MSE(DES)
0.5	0.3	468.14	10.43	4.30%	231.20
0.3	0.3	455.11	11.50	4.82%	247.24
0.7	0.5	480.39	10.19	4.24%	264.79

Figure 6.24 Product 2 DES in Minitab

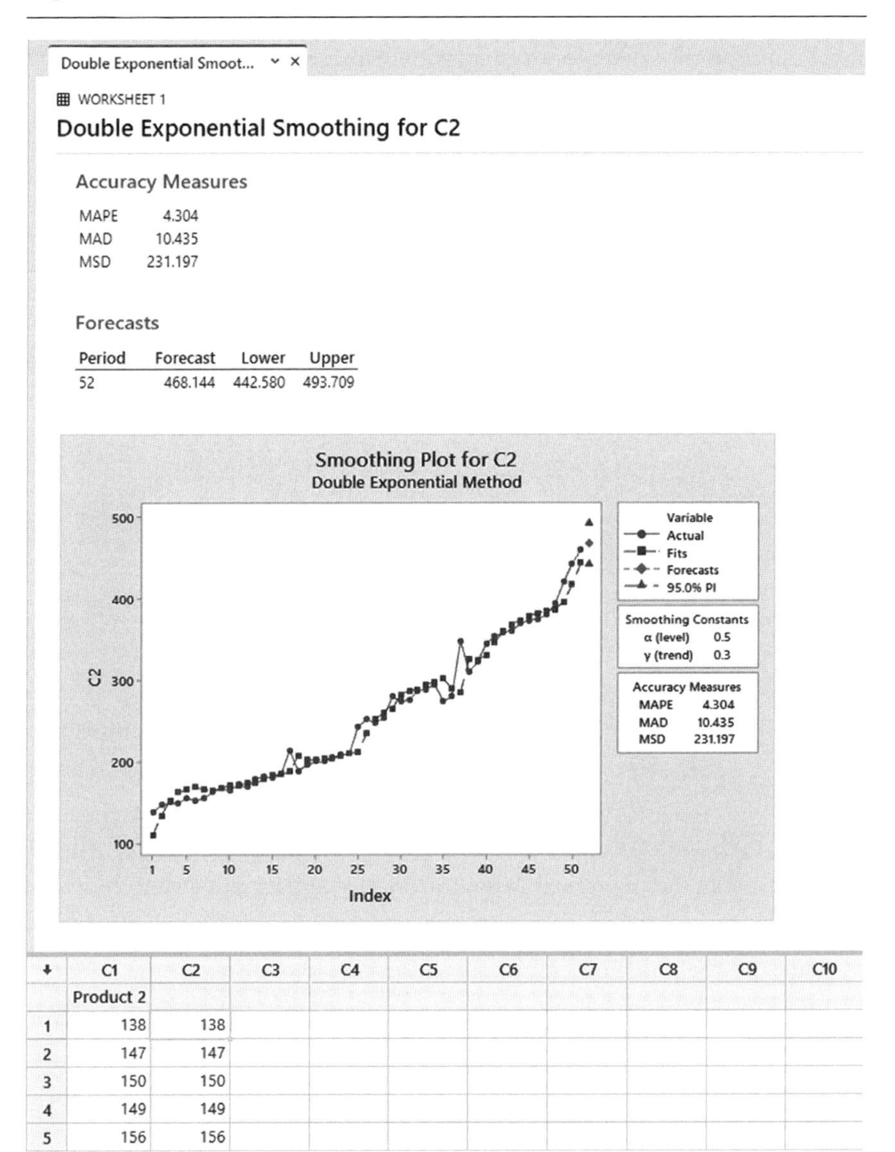

	C1	C2	C3	C4	C5	C6	C7	C8	C9	C10
	Product 2									
1	138	138								
2	147	147								
3	150	150								
4	149	149								
5	156	156								

Triple exponential smoothing

The triple exponential smoothing (TES) forecasting technique deals with data that presents clear seasonality. In this particular case, data can be analysed using three smoothing coefficients, such as α for the level in data, β for the trend in data and γ looking at the seasonality in the data provided.

There are two variations in dealing with seasonality, and this depends on the data behaviour. For data with stable seasonality, the additive model is more appropriate, where for the data with increase or decrease in seasonality the multiplicative model will be more suitable.

The additive model has the following representation:

$$F_{t+1} = L_t + k * T_t + S_{t-s+1}$$

Where the multiplicative model is:

$$F_{t+1} = (L_t + k * T_t) * S_{t-s+1}$$

The individual equations for the level (L_t), trend (T_t) and seasonality (S_t) are represented as follows:

$$L_t = \alpha * \frac{A_t}{S_{t-s}} + (1-\alpha) * (L_{t-1} + T_{t-1})$$

$$T_t = \beta * (L_t - L_{t-1}) + (1-\beta) * T_{t-1}$$

$$S_t = \gamma * \frac{A_t}{L_t} + (1-\gamma) * S_{t-s}$$

$F_t = L_t + T_t + S_{t-s}$, for the additive model and

$F_t = (L_t + T_t) * S_{t-s}$, for the multiplicative model

where:

A_t – represents the numerical value for the actual data at current period of time t

t – indicates the time now, so the current period of analysis

$t + 1$ – indicates time now plus one period ahead, therefore one period into the future

k – represents periods into the future, when associated with the trend, where $k = 1, 2, 3, \ldots$

s – is the assumed number of seasons in a cycle

L_t – is the level value at current time

L_{t-1} – is the level value at previous time period

T_t – is the value for trend at time now

T_{t-1} – is the trend value at the previous period of time

S_t – is the seasonal value at time t

S_{t-s} – is the seasonal value at time t–s, where s is the number of seasons within a cycle

S_{t-s+1} – is the seasonal value at time t–s+1 (in the future with one period from t–s)

F_t – is the forecast at period t. This is being calculated for all the values where actual data exists

F_{t+1} – is the value forecasted at period t+1

α – is the smoothing coefficient corresponding with the level, which ranges from 0 to 1

β – is the smoothing coefficient corresponding with the trend, which ranges from 0 to 1

γ – is the smoothing coefficient corresponding with seasonality, which ranges from 0 to 1

To carry out a forecasting analysis using the triple exponential smoothing method Steps 1, 2 and 3 identified for single exponential smoothing and double exponential smoothing are repeated in the case of triple exponential smoothing as:

Step 1. Identify the need to forecast.

Step 2. Determine the time period for which to forecast. This step requires the analysis to indicate for how many periods ahead is the forecast planned to take place. In the case of triple exponential smoothing, it is assumed that the forecast will be carried out for a minimum of one cycle.

Step 3. Identify and collect the data required to forecast. Clean and evaluate the collected data. An analysis of the data collected is expected in this case regarding the number of observations. This considers if all values over the observed period have been saved, if they are accurate, if there are any anomalies within the data collected.

Step 4. Evaluate whether TES would be the most appropriate forecasting technique to be used for the selected data. In this case there is the assumption that data has a level of noise, a trend and seasonality.

Step 5. Select the most appropriate value of alpha α, beta β and gamma γ coefficients for which the TES should take place. Values for $\alpha, \beta, \gamma \in (0, 1)$. For example, select the value for α, β and γ between 0.1, 0.3, 0.5, 0.7 and 0.9. Note the values selected for α, β and γ will not necessarily have the same value.

Step 6. Identify the L_0 by using the corresponding value from the trend equation. For T_0 use the corresponding value from the trend equation as well. Identify the value for s, the number of seasons in a cycle. If, for example, $s = 12$, for 12 months is a year, where the year is the cycle, values for S_0 to S_{12} need to be identified. They can follow the formula to $S_0 = A_0/L_0$ to $S_{12} = A_{12}/L_{12}$.

Step 7. In Excel, develop the template required for TES. Calculate the *level*, *trend* and *seasonal* factors for the first identified season the forecasting using the correct formulas. Calculate the forecasting using one of the two variations of the model (additive or multiplicative version). Calculate the forecasting errors (MAD, MAPE and MSE).

Step 8. Repeat Steps 5 to 7 for different values of alpha α, beta β and gamma γ.

Step 9. Construct the analysis table for different values of alpha α, beta β and L_0 gamma γ. Plot the data and the forecast on a graph for each calculated value of alpha α, beta β and gamma γ.

Step 10. Analyse the results and provide a full report regarding the use of this technique as well as the final forecasted results.

Analysis It has been identified that a forecasting analysis is required for the sales figures of product 3 for the following 12 months (Step 1 and Step 2). Sales data for product 3 has been identified and collected (Step 3) from the dedicated information system and is now ready for the analytics team to provide their evaluation. Trend and seasonality in data have been observed, therefore triple exponential smoothing (TES or Winter's model) has been selected for the analysis (Step 4).

Step 5. For the data collected, to start the analysis, a random selection for α, β and γ can be considered. These initially selected values can change as soon as the analysis is being developed. Let us assume that the initial values for α, β and γ are as follow: $\alpha = 0.7$, $\beta = 0.5$ and $\gamma = 0.3$.

Step 6. To be able to identify L_0, the trend equation is being used. As noted in Figure 6.25 the value for L_0 can be selected as 6561.3. The value for T_0 is now 111.9, and again this is selected from the trend equation. These two values for L_0 and T_0 can be considered for the first cycle, assuming that we have 12 seasons within a cycle. To calculate the value for S_0 to S_{12}, the formula to be used is A_0/L_0, A_1/L_{12}, … and A_{12}/A_{12}.

Step 7. The template developed in Excel for the TES is as indicated in Figure 6.25. The formula for level is inserted in position C15 and is represented by: =\$C\$1*(B15/E3) + (1-\$C\$1)*(C14+D14). In cell D15 the trend formula is inserted and this is: =\$D\$1*(C15-C14)+(1-\$D\$1)*D14 and in cell E15 the seasonality formula is inserted and this is: =\$E\$1*(B15/C15)+(1-\$E\$1)*E3.

Figure 6.25 Product 3 TES template

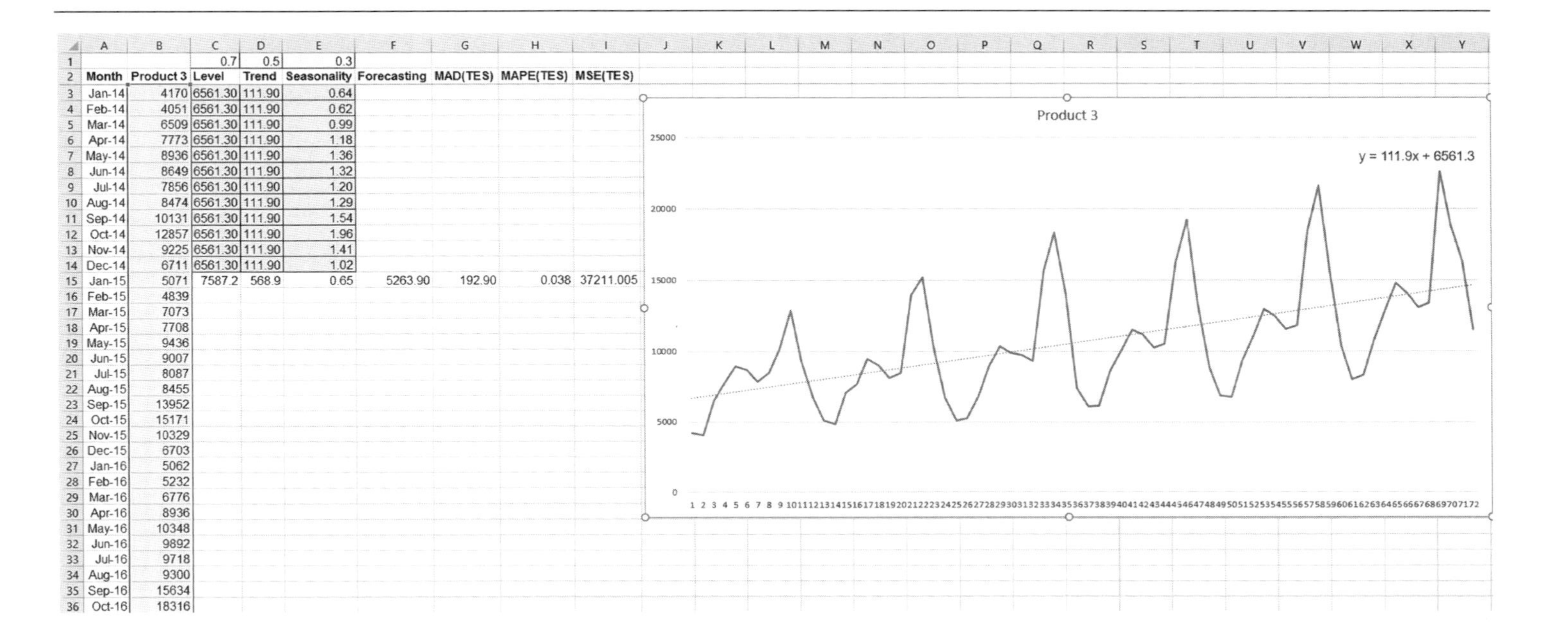

Month	Product 3	Level	Trend	Seasonality	Forecasting	MAD(TES)	MAPE(TES)	MSE(TES)
		0.7	0.5	0.3				
Jan-14	4170	6561.30	111.90	0.64				
Feb-14	4051	6561.30	111.90	0.62				
Mar-14	6509	6561.30	111.90	0.99				
Apr-14	7773	6561.30	111.90	1.18				
May-14	8936	6561.30	111.90	1.36				
Jun-14	8649	6561.30	111.90	1.32				
Jul-14	7856	6561.30	111.90	1.20				
Aug-14	8474	6561.30	111.90	1.29				
Sep-14	10131	6561.30	111.90	1.54				
Oct-14	12857	6561.30	111.90	1.96				
Nov-14	9225	6561.30	111.90	1.41				
Dec-14	6711	6561.30	111.90	1.02				
Jan-15	5071	7587.2	568.9	0.65	5263.90	192.90	0.038	37211.005
Feb-15	4839							
Mar-15	7073							
Apr-15	7708							
May-15	9436							
Jun-15	9007							
Jul-15	8087							
Aug-15	8455							
Sep-15	13952							
Oct-15	15171							
Nov-15	10329							
Dec-15	6703							
Jan-16	5062							
Feb-16	5232							
Mar-16	6776							
Apr-16	8936							
May-16	10348							
Jun-16	9892							
Jul-16	9718							
Aug-16	9300							
Sep-16	15634							
Oct-16	18316							

To calculate the forecasting for the first period, the multiplicative version of the formula can be considered. In cell F15 this can be inserted as: =(C15+D15)*E15.

The forecasting errors will be calculated next. For MAD(TES), is cell G15 =ABS(B15-F15); MAPE(TES) is in cell H15=ABS((B15-F15)/B15) and the MSE(TES) in cell I15 with the formula: =(B15-F15)^2.

At this point the evaluation is as in Figure 6.25.

Completing this analysis, the results are presented as indicated in Figure 6.26. The formula used in cell F75 is: =($C$74+1*$D$74)*E63. In cell G75 we have the completed formula to calculate the MAD as: =AVERAGE(G15:G74). We also have average formulas in cells H75 and I75 to be able to calculate the MAPE and MSE. The results indicated for the randomly selected values for α, β and γindicate values such as MAD = 316.06; MAPE = 2.64% and MSE = 187395.05. These are to be considered relatively low values, however at this point we are not yet in a position to provide a full analysis until different values for the selected coefficients are being considered.

Step 8. Different values for the coefficients can be considered at this point. For example, the α coefficient can be increased to 0.8, β decreased to 0.3 and γ increased slightly to 0.5. If all the values are recalculated all the errors values are decreased such as for the MAD(TES) = 201.06, MAPE(TES) = 1.69% and MAPE(TES) = 78102.99. This indicates that the direction of modifying the coefficients is to follow this trend, where the coefficients can be further adjusted.

There is one other method that can be used to identify the best combination for the selection of the smoothing coefficients α, β and γ when Excel add-in Solver software is being used.

Solver is selected from the Data tab in Excel. There is further instruction on how to install Solver later on in Chapter 7. For the set objectives, one of the forecasting error can be selected such as MAD(TES); this is to be minimized; where for the changing variables cells the three coefficients are to be selected. As constraints, for each of these coefficients the conditions to be >= with 0.1 and <= with 0.9 are considered. Using Solver this way, returns a set of values such as 0.9 for alpha, 0.1 for beta and 0.64 for gamma.

Step 9. The analysis table is to be constructed at this point with all the selected values for the coefficients, where the forecasted values for all three sets of data to be represented within a graph. These are now represented in Figure 6.27.

Step 10. It can be noted that the lowest values for the MAD, MAPE and MSE are obtained for the set 3 of the analysis carried out for this forecasting evaluation.

Figure 6.26 Product 3 TES calculations

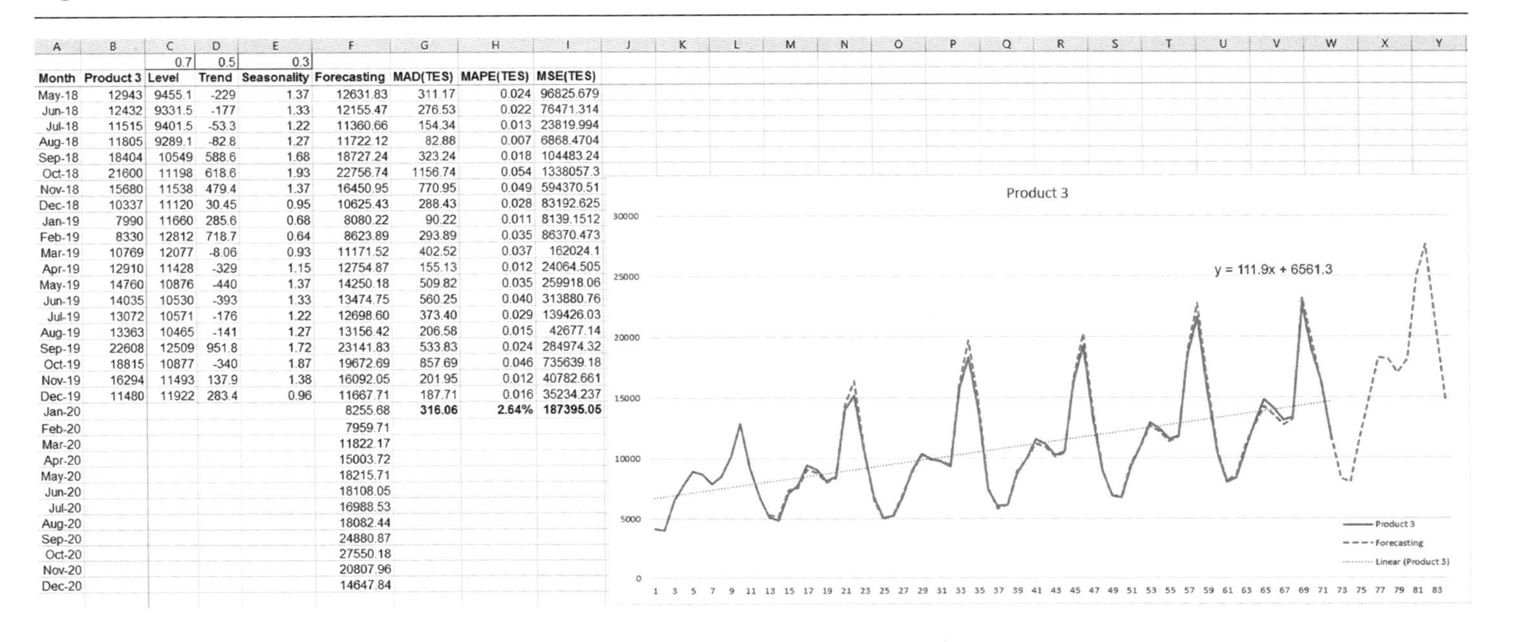

		0.7	0.5	0.3				
Month	**Product 3**	**Level**	**Trend**	**Seasonality**	**Forecasting**	**MAD(TES)**	**MAPE(TES)**	**MSE(TES)**
May-18	12943	9455.1	-229	1.37	12631.83	311.17	0.024	96825.679
Jun-18	12432	9331.5	-177	1.33	12155.47	276.53	0.022	76471.314
Jul-18	11515	9401.5	-53.3	1.22	11360.66	154.34	0.013	23819.994
Aug-18	11805	9289.1	-82.8	1.27	11722.12	82.88	0.007	6868.4704
Sep-18	18404	10549	588.6	1.68	18727.24	323.24	0.018	104483.24
Oct-18	21600	11198	618.6	1.93	22756.74	1156.74	0.054	1338057.3
Nov-18	15680	11538	479.4	1.37	16450.95	770.95	0.049	594370.51
Dec-18	10337	11120	30.45	0.95	10625.43	288.43	0.028	83192.625
Jan-19	7990	11660	285.6	0.68	8080.22	90.22	0.011	8139.1512
Feb-19	8330	12812	718.7	0.64	8623.89	293.89	0.035	86370.473
Mar-19	10769	12077	-8.06	0.93	11171.52	402.52	0.037	162024.1
Apr-19	12910	11428	-329	1.15	12754.87	155.13	0.012	24064.505
May-19	14760	10876	-440	1.37	14250.18	509.82	0.035	259918.06
Jun-19	14035	10530	-393	1.33	13474.75	560.25	0.040	313880.76
Jul-19	13072	10571	-176	1.22	12698.60	373.40	0.029	139426.03
Aug-19	13363	10465	-141	1.27	13156.42	206.58	0.015	42677.14
Sep-19	22608	12509	951.8	1.72	23141.83	533.83	0.024	284974.32
Oct-19	18815	10877	-340	1.87	19672.69	857.69	0.046	735639.18
Nov-19	16294	11493	137.9	1.38	16092.05	201.95	0.012	40782.661
Dec-19	11480	11922	283.4	0.96	11667.71	187.71	0.016	35234.237
Jan-20					8255.68	**316.06**	**2.64%**	**187395.05**
Feb-20					7959.71			
Mar-20					11822.17			
Apr-20					15003.72			
May-20					18215.71			
Jun-20					18108.05			
Jul-20					16988.53			
Aug-20					18082.44			
Sep-20					24880.87			
Oct-20					27550.18			
Nov-20					20807.96			
Dec-20					14647.84			

Figure 6.27 Product 3 TES with analysis table

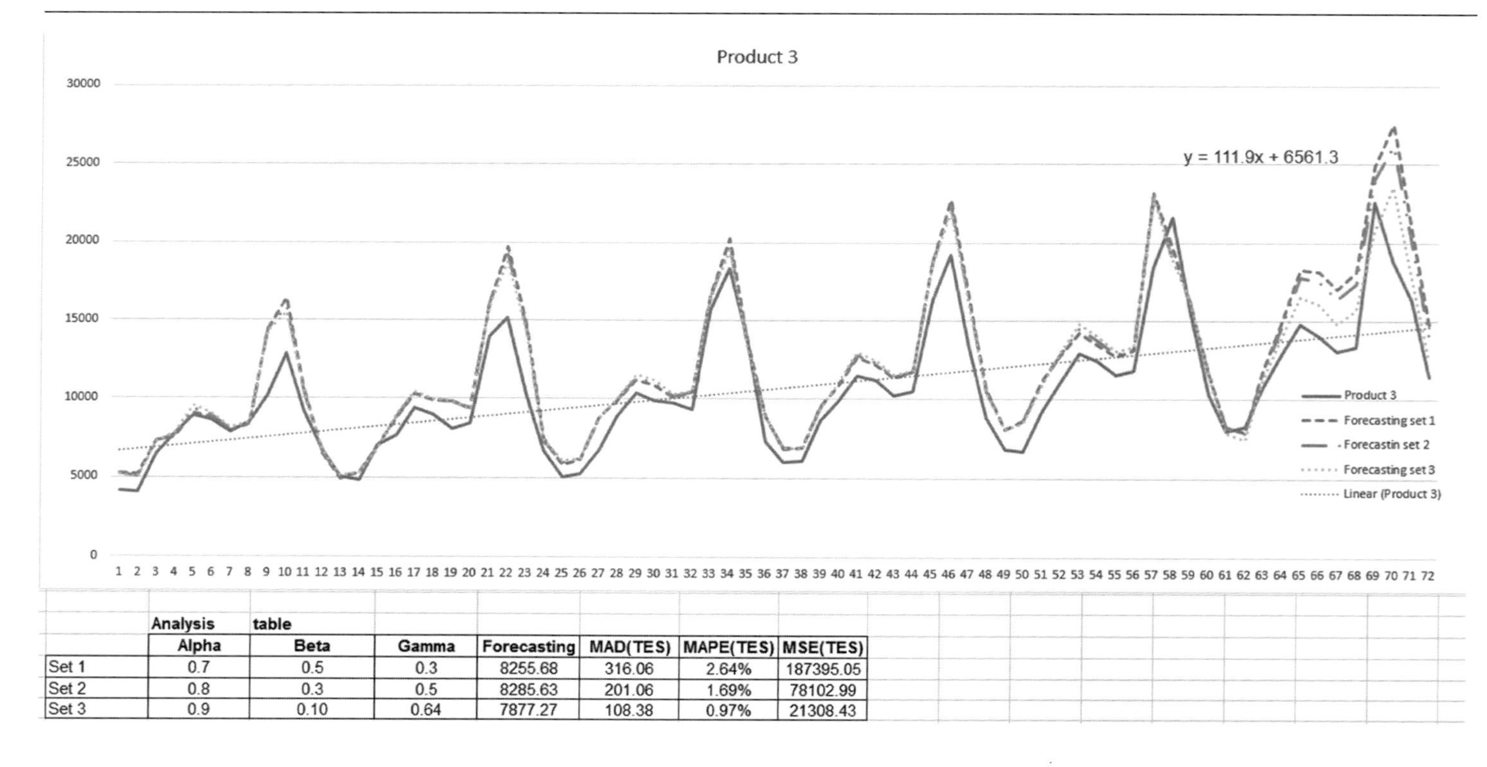

Analysis table

	Alpha	Beta	Gamma	Forecasting	MAD(TES)	MAPE(TES)	MSE(TES)
Set 1	0.7	0.5	0.3	8255.68	316.06	2.64%	187395.05
Set 2	0.8	0.3	0.5	8285.63	201.06	1.69%	78102.99
Set 3	0.9	0.10	0.64	7877.27	108.38	0.97%	21308.43

Figure 6.28 Product 3 TES in Minitab

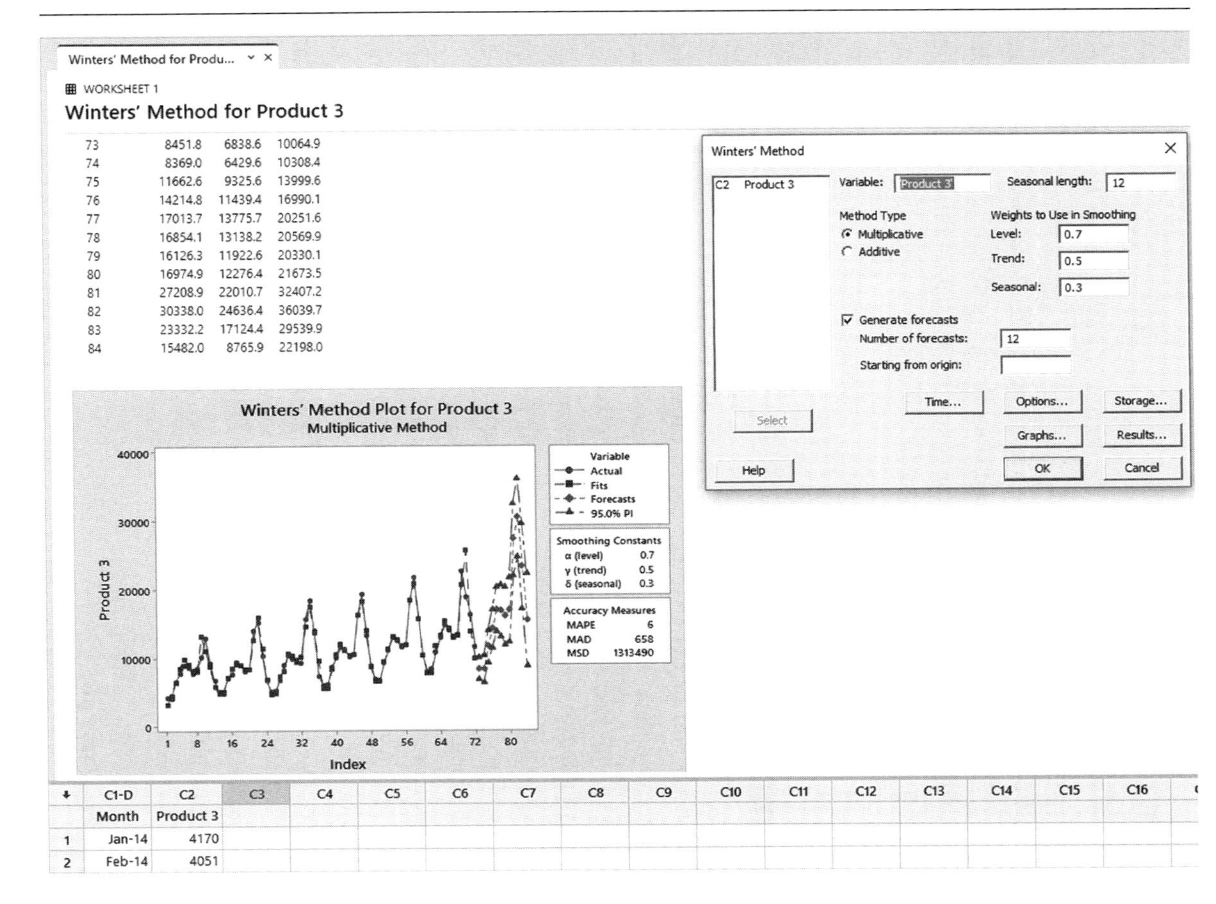

Triple exponential smoothing analysis using Minitab 19

For the sales data for Product 3 inserted in the spreadsheet section of the Minitab software as in Figure 6.28, the instructions to follow to carry out a triple exponential smoothing analysis are:

Stat > Time Series > Winter's Method

A new screen appears, where for the section *Variable*, select *Product 3* and for the section *Seasonal Length* use *12*.

The results obtained in this case are slightly different with those obtained in Excel for the same values of the smoothing coefficients α, β and γ. The calculated values for the forecasting analysis are slightly different, and this is due to not having full visibility on the assumptions used when considering the analysis in Minitab.

Summary

The main aim of the chapter was to capture models of a descriptive and predictive nature that are used in the area of logistics and supply chain with the note that they can be used in many other areas of operations.

A number of descriptive models have been detailed, together with the approach to use and how to use them using particular software packages such as Excel and Minitab. Many other descriptive models could be considered, but the focus has been on some of the most popular models.

Predictive models have also been detailed with particular examples used in each case. A particular focus was given to the models used for forecasting, this being one of the key areas that contributes to the operation and strategic planning within a supply chain. The theoretical aspect of the forecasting models has been detailed, where individual examples show how these examples can be used in practice. Again, these models have been detailed using templates developed in Excel and instructions were given on how to operate with these using Minitab.

References

Albright, S C and Winston, W L (2011) *Management Science Modeling*, 4th edn, South-Western College Pub, Andover and Mason, OH

Arunraj, N S, and Ahrens, D (2015) A hybrid seasonal autoregressive integrated moving average and quantile regression for daily food sales forecasting, *International Journal of Production Economics*, 170, 321–35, doi:10.1016/j.ijpe.2015.09.039

Baker, K R (2015) *Optimization Modeling with Spreadsheets*, 3rd edn, Wiley, Hoboken, NJ

Beamon, B M, and Chen, V C P (2001) Performance analysis of conjoined supply chains, *International Journal of Production Research*, 39(14), 3195–218

Curwin, J and Slater, R (2004) *Quantitative Methods*, Thomson, London

Curwin, J and Slater, R (2008) *Quantitative Methods for Business Decisions*, Thomson Learning, London

Curwin, J, Slater, R and Eadson, D (2013) *Quantitative Methods for Business Decisions*, 7th edn, Cengage Learning, Andover

El-Sakty, K, Tipi, N, Hubbard, N and Okorie, C (2014) The development of a port performance measurement system using time, revenue and flexibility measures, *International Journal of Business and General Management*, 3(4), 17–36

Evans, J R (2016) *Business Analytics: Methods, models, and decisions*, 2nd edn, Pearson, Boston, MA

Field, A P (2018) *Discovering Statistics using IBM SPSS Statistics*, 5th edn, SAGE, Los Angeles

Kubiak, T M and Benbow, D W (2016) *The Certified Six Sigma Black Belt Handbook*, 3rd edn, ASQ Quality Press, Milwaukee, MI

Lepenioti, K, Bousdekis, A, Apostolou, D and Mentzas, G (2020) Prescriptive analytics: Literature review and research challenges, *International Journal of Information Management*, 50, 57–70, doi:10.1016/j.ijinfomgt.2019.04.003

Merkuryeva, G, Valberga, A and Smirnov, A (2019) Demand forecasting in pharmaceutical supply chains: A case study, *Procedia Computer Science*, 149, 3–10, doi:10.1016/j.procs.2019.01.100

Mumtaz, U, Ali, Y and Petrillo, A (2018) A linear regression approach to evaluate the green supply chain management impact on industrial organizational performance, *Science of the Total Environment*, 624, 162–69, doi:10.1016/j.scitotenv.2017.12.089

Bibliography

Anderson, D R (2016) *Quantitative Methods for Business*, 13th edn, Cengage Learning, Boston, MA

Balakrishnan, N (2017) *Managerial Decision Modeling: Business analytics with spreadsheets*, 4th edn, DeG Press, Boston, MA

Maisel, L and Cokins, G I (2014) *Predictive Business Analytics: Forward-looking capabilities to improve business performance*, John Wiley & Sons, Inc, Hoboken, NJ

Ragsdale, C T (2018) *Spreadsheet Modeling and Decision Analysis: A practical introduction to business analytics*, 8th edn, Cengage Learning, Boston, MA

Waters, C D J (2011) *Quantitative Methods for Business*, 5th edn, Financial Times/Prentice Hall, Harlow

Supply chain analytics 07

Prescriptive models

LEARNING OBJECTIVES

- Be aware of typical prescriptive models used in the area of business analytics with specific application to the supply chain system.
- Review a set of prescriptive BA models such as resource allocation, covering, transportation, transshipment models and algorithms used to solve vehicle routing problems.
- Consider in particular the required input data and the expected output data for each set of models.
- Understand how to develop and analyse prescriptive models with the use of spreadsheets.

Introduction

Supply chain analytics could have applications in the supply chain systems in all areas of the supply chain management functions. For example, in the sourcing and procurement areas analytics could be used to identify the best suppliers or best resources to use to procure materials and to identify the best routes to collect semi-finished products to be brought in-house. Analytics can also be used in the distribution area, to provide the best distribution plan, in the supply chain network design, in product design, in production planning in finance, in quality control and so on.

In a systematic literature review carried out from 2010 to 2019, Lepenioti et al (2020) have been able to identify a number of prescriptive models, which they classify in the following categories: probabilistic models, machine learning/data mining, mathematical programming models, evolutionary computation models, simulation models and logic-based models. These models have been noted to have applicability in other areas than logistics and supply chain. However, the list of identified models is interesting to note as reference:

- Probabilistic: Markov decision process, hidden Markov model and Markov chain.
- Machine learning/data mining: k-means clustering, reinforcement learning, privacy preservation, Boltzmann machine, Nadarays-Watson estimation, and artificial neural networks.
- Mathematical programming: Mixed integer programming, linear programming, binary quadratic programming, non-liner programming, binary linear integer programming, stochastic optimization, conditional stochastic optimization, constrained Bayesian optimization, fuzzy linear optimization, robust and adaptive optimization, dynamic programming and optimal search path.
- Evolutionary computation: Genetic algorithm, evolutionary optimization, greedy algorithm, particle swarm optimization.
- Simulation: Simulation under random forest, risk assessment, stochastic simulation and what-if scenarios. Other simulation models could also be considered, but they may have not been captured under the distinct prescriptive analytics description of this review. These may be discrete event simulation, discrete simulation and dynamic simulation models.
- Logic-based models: Association rules, decision rules, criteria-based rules, fuzzy rules, distributed rules, benchmark rules, desirability function, graph-based recommendation and 5W1H (Lepenioti et al, 2020).

However, although we note the list of models captured in this study is comprehensive, other models may be referenced when dealing with specific aspects of supply chain analysis.

A number of models have been developed and published that are relevant to the supply chain and deal with supply chain issues. Along the supply chain, it can be observed that different analytical models are applied at different levels and with various degree of complexity. For example, forecasting could be employed at product and organization level, followed by a similar forecasting function at a different echelon in the chain for a number of different products that is then taken to a planning/tactical level for that particular organization. Knowledge learned from a forecasting analysis at a particular level within an organization could bring significant benefits to another organization within the same supply chain. The key here will be the level of knowledge/information integration and sharing between the two organizations.

Optimization models using linear programming

As explained in the earlier chapters, prescriptive models are those that help us to understand what can happen, and what can be done about this. Some of the models detailed in this chapter will mainly focus on resources, capacity allocation, covering, assignment, direct transportation models, transshipment and vehicle routing problems. However, this list is not exhaustive, and other models have been considered and implemented in supply chains. Some of the models considered in this chapter are solved with the use of linear programming, optimization approach using Excel with the main emphasis on their role in the supply chain. These models are known as *deterministic*, as the input data and the other parameters considered for them are known with certainty (Balakrishnan, 2017). Particular examples are constructed with the use of linear equations and linear inequalities for each of the each of these models; however, the key benefit coming from the discussion is provided in relation to their application to the supply chain. Similar models are present in materials in the area of management science, operational research and operations management, whereas in this book the main contribution is made to their application in the field of supply chain.

Characteristics of linear programming models

Models using the linear programming technique have the following characteristics:

- **Variables:** The elements that we try to identify within a linear programming model are known as variables. These could be *number of products to produce*, or *number of products to transport*, or *number of workers required per shift* and so on. Defining the variables is a key step in solving linear programming models and they form part of the mathematical model.
- **Constraints:** The set limitations within which the objective function of a model can be reached are known as constraints. It is particularly important to include the full list of constraints when aiming to solve a linear programming model. However, some of these may not be incorporated if they do not significantly affect the final outcome. There are nonnegativity constraints, where the variables we look to identify are >= 0. Constrains could also consider integer constraints, where the variables are expected to be 'full' values. For example, in solving a linear programming model that aims to produce a chair, the integer constraint will ensure that no

partial chairs are being produced. There are also constraints that will be identified as inequality constraints.

- **Objective function:** These models are characterized by having an objective function. In general, this is set to maximize for example 'profit' or minimize for example 'cost'. This function needs to be clearly defined as a mathematical function. When only one objective function is defined, the model is known as a single objective function model. However, having only one objective function model doesn't imply that the model only deals with one performance measure. Within a single objective function, more than one performance measure could be set to characterize the model (such as cost, deliver in full, resource utilization, available inventory and others). There are also mathematical models that operate with multiple objective functions, where more than one objective is developed and optimized mathematically.

Resource allocation models

Resource allocation models aim to maximize their objective by meeting resource constraints. For this type of model, the resources are divided and assigned to competing activities (Baker, 2016; Powell and Baker, 2011). This model has the potential to consider more than one dimension within a supply chain. It can incorporate details about the manufacturing process, the volume, the number of products to be produced, the quantity of raw materials or semi-finished products supplied, and could also incorporate details about customer demand. However, although the model works only with one optimization function (maximize profit), more than one performance measure can be observed in this context, such as: maximum profit, throughput, resource availability and resource utilization and customer service.

To understand this type of model in the context of supply chain, the following two examples are presented, analysed and discussed: the resource allocation example and the extended resource allocation example. These two examples follow a similar setting, with the extended example introducing the notion of time, where the analysis can be carried out over a period of time. This example is manly analysed from a manufacturing point of view, where information and data are available to and from the suppliers and customers.

Resource allocation example

A manufacturing organization aims to bring further understanding to its operations with specific reference to two of its key products (namely NT111 and NT112). The company has identified a set of issues with these two products and aims to gain further understanding from analysing these using prescriptive analytics.

Identified issues

- There are issues concerning the available time allocated for producing these two products in the manufacturing and assembly department.
- There are issues identified with raw materials available in stock from two suppliers.

Many other issues can be considered for analysis at this stage, but the two main aspects that are of concern are those presented above. The analysis is intended from the point of view of the total profit generated by producing these two products, considering the issues identified above. By producing these two products, the company is able to provide an estimate of the profit per product as well as an overall profit.

Before the analysis can start, the first point for consideration is regarding data. In general, this is data that is available internally within the organization's IT system, or needs to be collected by an analysis of internal sources.

Input data. To start the analysis the following input data is required:

- **Raw material required per product.** This type of data is a product design characteristic and is indicated at the product design stage, and it will be present in the bill of materials that forms this product. It is important to realize that this is a detail that is not changeable. As soon as changes are to be considered to this component, the overall product characteristics change. Therefore, changes in product design and changes in the characteristics of its components will result into a new product.
- **Total number of units sourced from suppliers.** This type of data is available from the sourcing/buying department. This is a value that can change for a number of reasons, some of which may be: the stock availability at the supplier end, or at the manufacturer end, the delivery conditions and the quality of the raw material supplied. As this information is linked to the stock availability, inventory control managers also deal with this type of information.
- **Time required to produce the indicated products.** This type of information is part of the product characteristics, identified through the design process that can, for example, clearly specify the time require for a product to be processed at a particular machine. This type of data is unchangeable, and

will be captured in the bill of materials for each product. This is relevant to the planners, manufacturers and operations managers. However, there may be an added component to this time that may consider waiting time in the queue to be processed that potentially can be reduced with a more optimal operation. This will be a dynamic component, and however relevant it is to consider this aspect, for the purpose of this example it will not be taken into consideration. Only a static component is considered.

- **Total time allocated for producing these two products.** This information is provided by the production planner, and depends on the resources available in the manufacturing departments identified. The total number of hours allocated per product in manufacturing is information available from the planning manager. This value can change due to staff and/or machine availability. However, other issues could be encountered at this point when unpredictable disturbances occur, for example, with machine downtime, or accidents, or power cut and others.
- **Profit per product.** This detail is available from the finance department within an organization and is in general based on past information as well as details acquired from current market conditions.

These details may be readily available within the ERP system; however, time may be allocated to allow for the collection of this information. These details are presented in Figure 7.1.

Figure 7.1 Product allocation – initial data

Products	NT111	NT112	
Quantity of products to be produced			
Unique product characteristics			*Resources availability*
Raw material 1 from supplier 1	4	5	450
Raw material 2 from supplier 2	3	3	300
Time in manufacturing department (min)	20	25	2000
Time in the assembly department (min)	15	30	1800
Profit per product (£)	5	7	0

Scope. The scope of the analysis is threefold:

- Identify the *total number of products to be produced.*
- Maximize the *profit* to produce these two products.
- Maximize the *use of the resources* put in place for producing these products.

Output data. The output data obtained at the end of the analysis is as follows:

- *Quantity to be produced per product.*
- *Quantity of raw material used in production.*
- *Leftover inventory of raw materials supplied.*
- *Overall time spend in production per department.*
- *Resource utilization.*
- *Total profit.*

Therefore, in this particular example, we are dealing with fixed and variable data as well as data that is required from different locations and functions within a company.

Therefore, the performance measures that can be identified within this model are as presented in Table 7.1.

As soon as the data collection process has been completed, the analysis can continue by employing a linear programming optimization approach. To work with a linear programming approach, three elements are to be

Table 7.1 Performance measures in the allocation model

Performance measure	Type of measure
Total number of products to be produced	Quantity measure
Quantity produced per product	Quantity measure
Quantity of raw material used in production	Quantity measure
Leftover inventory for each raw material	Quantity measure
Total profit	Financial measure
Utilization of resources	Asset utilization measure
Time spent in production per department	Time-related measure

defined: the variables, the constraints and the objective function by which we aim to evaluate a situation. This will form the mathematical model as detailed below.

Variables. Variables are things that we don't know and we try to find out. Following these particular examples, the variables are *the total number of products to be manufactured*, therefore the total number of NT111 and NT112. We may want to represent these using mathematical notations such as x and y; however, for this example we will avoid using these notations.

- **Variable 1:** Total number of products to be product NT111.
- **Variable 2:** Total number of products to be product NT112.

Constraints. In this particular case there is a set of limitations that need to be considered. These are supplier-related constraints named raw material 1 and raw material 2. There are manufacturing dependant constrains named manufacturing time and assembly time:

- **Raw material 1:** The total quantity of raw material 1 required to produce the two products is less than or equal to (<=) the quantity available in stock for raw material 1.
- **Raw material 2:** The total quantity of raw material 2 required to produce the two products is less than or equal to (<=) the quantity available in stock for raw material 2.
- **Manufacturing time:** The total time required in the manufacturing department to produce the two products is less than or equal to (<=) the total allocated time.
- **Assembly time:** The total time required in the assembly department to produce the two products is less than or equal to (<=) the total allocated time.

To work with the linear programming representation, two other types of constraints are to be considered. One is the non-negative constraint that refers to set the two variables to higher than or equal to value '0'. This implies that we should not expect negative values as part of the solution.

- *Variable 1; Variable 2 >= 0*

There is also the expectation that we will have 'integers' value. In other words, we will expect to work with full products.

- *Variable 1; Variable 2 = integer*

Objective function. To complete the setting of our linear programming model, the objective function is required. In this case we will have as an objective the *total profit* that we will want to maximize.

Total profit = maximize (total profit)

The mathematical model is now set.

This model can be implemented and solved in Excel using the add-in software Solver. Considering the data set as presented in Figure 7.1, based on the input data collected a template can be developed in Excel that captures all the required elements, as shown in Figure 7.2.

Solving this example in Excel, the following formulas need to be considered for each of the constraints set.

In cell F5 input =SUMPRODUCT(D5:E5,$D$2:$E$2)

In cell F6 input =SUMPRODUCT(D6:E6,$D$2:$E$2)

In cell F7 input =SUMPRODUCT(D7:E7,$D$2:$E$2)

In cell F8 input =SUMPRODUCT(D8:E8,$D$2:$E$2)

The value for the *objective function* to maximum profit will be listed in cell H10. This will have the following formula =SUMPRODUCT(D10:E10, D2:E2).

Having set up all the formulas in the template, the model is now ready to be solved using 'Solver'. Solver is an add-on programme that will need to be set up in Excel. The following instructions could be followed to set this up:

In Excel select *File-> Options-> Add-ins-> Solver Add-in* and select Go.

Figure 7.2 Resource allocation model – template

	A	B	C	D	E	F	G	H	I
1	Variables		Products	NT111	NT112				
2			*Quantity of products to be produced*						
3									
4			*Unique product characteristics*			*Used in production*		*Resources availability*	*Left over after production*
5	Constraints		Raw material 1 from supplier 1	4	5	0	<=	450	450
6			Raw material 2 from supplier 2	3	3	0	<=	300	300
7			Time in manufacturing department (min)	20	25	0	<=	2000	2000
8			Time in the assembly department (min)	15	30	0	<=	1800	1800
9									
10	Objective		Profit per product (£)	5	7			0	

The *Solver* will appear under the *Data* tab in Excel. As presented in Figure 7.3, individual setting for the variables, constraints and the objective function can be set.

Variables. Variable 1 is under cell D2 and Variable 2 is under cell E2. Therefore, in Solver, these will be inputted in the section *By Changing Variable Cells* as $D$2:$E$2.

Figure 7.3 Resource allocation model – Solver

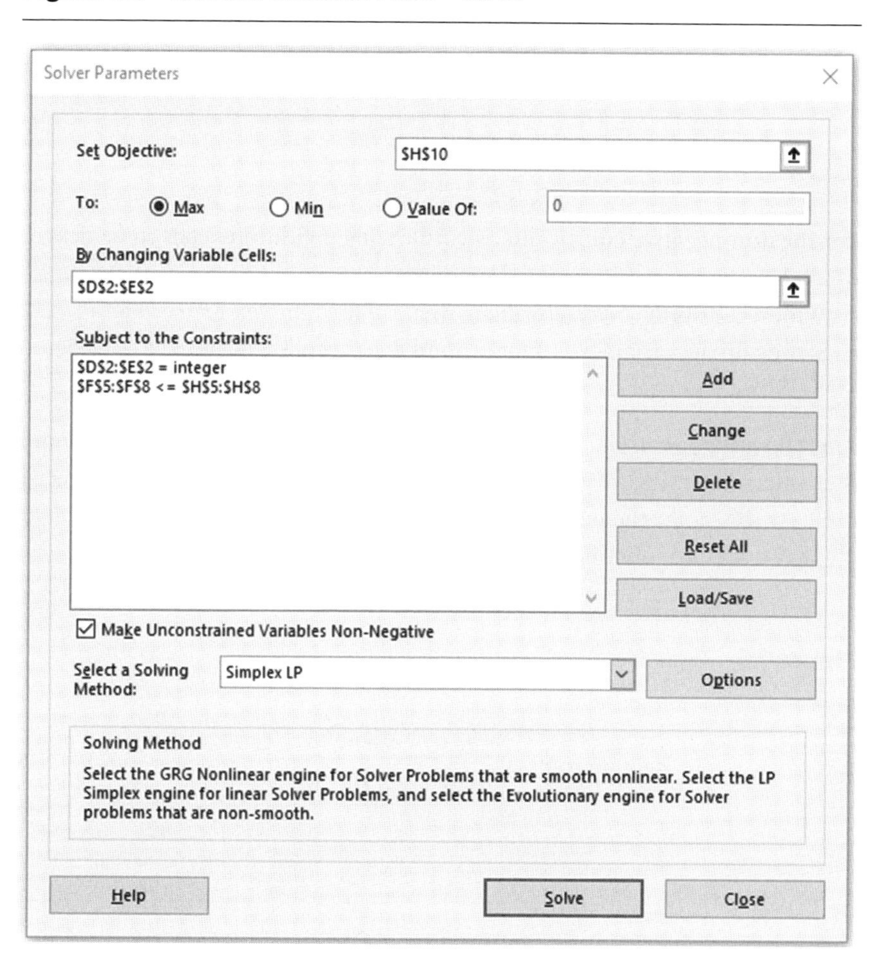

Constraints. The raw materials and time to produce constraints in this case are in cells F5, F6, F7 and F8. In Solver you need to press the *Add* button to open the add constraints values, as indicated in Figure 7.4.

Figure 7.4 Resource allocation model – adding constraints in Solver

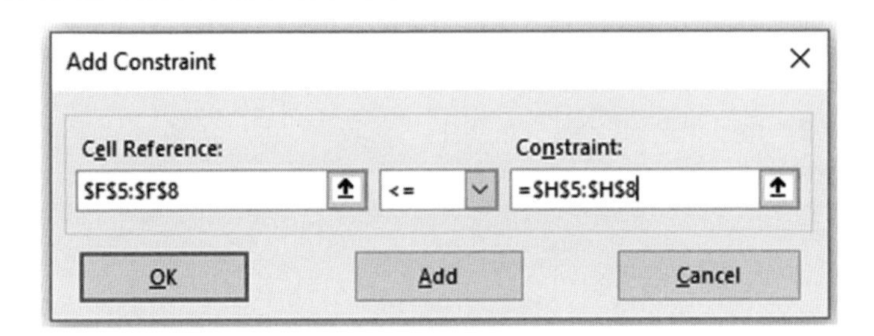

For the non-negative constraints, the following condition needs to be ticked: *Mark Unconstrained Variables Non-Negative.*

Within the option *Select a Solving Method* use Simplex LP (see Figure 7.3).

Setting up the variables as integer numbers, the following is to be considered $D$2:$E$2 = integer.

The objective function. In Solver in the *Set Objectives*, select $H$10 and then select Max (Figure 7.3). This will maximize the function inserted in cell H10. At this point the model will be ready to run Solver by selecting *Solve*.

The results are obtained as presented in Figure 7.5.

Figure 7.5 Resource allocation model – solution

	A	B	C	D	E	F	G	H	I
1	Variables		Products	NT111	NT112				
2			*Quantity of products to be produced*	66	27				
3									
4			*Unique product characteristics*			*Used in production*		*Resources availability*	*Left over after production*
5	Constraints		Raw material 1 from supplier 1	4	5	399	<=	450	51
6			Raw material 2 from supplier 2	3	3	279	<=	300	21
7			Time in manufacturing department (min)	20	25	1995	<=	2000	5
8			Time in the assembly department (min)	15	30	1800	<=	1800	0
9									
10	Objective		Profit per product (£)	5	7			519	

The data obtained as a result of solving this model is *Quantity to be produced per product*. It can be noted that the results for each product are listed in cells D2 and E2 respectively.

Quantity to be produced for product NT111= 66 units and

Quantity to be produced for product NT112= 27 units.

Therefore, the *total quantity to be produced* is 93 units in total.

Quantity of raw material used in production. The results in this case are listed under the cells F5 and F6. It can be noted that the quantity used in production is less than or equal to the quantity available in stock. From raw material 1 only 399 units have been used of 450 available and for raw material 2 only 279 units have been used of the 300 available. Still, at this point there is material left over. The leftover quantity is listed in cell I5 for raw material 1 and I6 for raw material 2.

Table 7.2 Performance measures – results

Performance measure	Value
Total number of products to be produced	93 units
Quantity produced per product 1	66 units
Quantity produced per product 2	27 units
Quantity of raw material 1 used in production	399 units
Quantity of raw material 2 used in production	279 units
Leftover inventory for raw material 1	51 units
Leftover inventory for raw material 2	21 units
Total profit	£519
Resources utilized in the manufacturing department	99.75%
Resources utilized in the assembly department	100%
Time spent in manufacturing department	1995 min
Time spent in manufacturing department	1800 min

Overall time spent in production per department. The results are listed under the cells F7 and F8. It can be noted that the time spent is less than or equal to the time available, and at this point there is some time left (5 min, see cell I7) in the manufacturing department, but not in the assembly department.

Total profit. The total profit is listed in cell H10 with a value of £519.

A response to the set of performance measures for this model can now be noted in Table 7.2.

To make best use of this model for decision-making, a number of 'What if...?' scenarios could be considered for the analysis. These scenarios may include:

Observations. (1) What if the stock available for both raw materials increases? Let's assume as an example that the values will be double from the original setting. Without any change in the time allocated for producing these products (time in the manufacturing and assembly department), it can be noted that there will be no changes in the results obtained after running the Solver again.

(2) One other scenario could be linked to customer demand. What if the demand for product 2 is set to 40 units? This change can be inserted as one of the constraints in the system, however it will produce results as indicated in Figure 7.6.

Figure 7.6 Resource allocation – adding demand constraints

	A	B	C	D	E	F	G	H	I
1	Variables		Products	NT111	NT112				
2			*Quantity of products to be produced*	40	40				
3									
4			*Unique product characteristics*			*Used in production*		*Resources availability*	*Left over after production*
5	Constraints		Raw material 1 from supplier 1	4	5	360	<=	450	90
6			Raw material 2 from supplier 2	3	3	240	<=	300	60
7			Time in manufacturing department (min)	20	25	1800	<=	2000	200
8			Time in the assembly department (min)	15	30	1800	<=	1800	0
9									
10	Objective		Profit per product (£)	5	7			480	

In this case, it can be observed that the production is better balanced, but there are more materials left in stock as well as more time unused in the manufacturing department. However, the profit in this case is now lower.

In conclusion, it can be noted that although this tool is mainly used for optimization, it can also be used for prediction. Setting up a model will allow understanding on a number of performances, as indicated in Table 7.2.

A number of other scenarios can be considered in the case of this model, such as different demand per product, where the performance measure will be to meet the customer demand, not necessarily to maximize throughput, or maximize profit. Similar examples are detailed in publications such as Baker (2016, p 29) with an example on Brown Furniture Company discussing the allocation of three products.

There is the opportunity for this type of model to be used for prediction and to further enhance the decision-making process for managers in the supply chain. Therefore, an example indicating an extended version of the resource allocation example is detailed next.

Extended resource allocation example

The same manufacturing organization is looking to analyse three of their products (NT121, NT122, NT123) over a period of three months. After these three months, three more products are to be introduced into the analysis and the analysis is to continue for another three months with now six products in total (NT121, NT122, NT123, NT131, NT132 and NT133).

The raw material supplied from two more suppliers needs to be considered in the second phase of production for products NT131, NT132 and NT133.

The time allocated in the manufacturing and assembly department is considered the same for three consecutive months. Additional time is allocated in the manufacturing and assembly department during phase two of the analysis to take into consideration the introduction of the three new products.

The reorder quantity for each raw material considered within this analysis has been calculated by the inventory control manager and the same order for raw materials is to be placed at the beginning of each month. Due to the raw material characteristics, the material left in stock after production can be reused during the following month's production.

Details about the estimated profit for each product are also provided.

Input data. A similar structure for the input data is followed as explained in the previous example. Similar conditions are followed regarding the *raw material required per product*; *total number of units sourced from suppliers* for all three products; *time required to produce the indicated products* in both the manufacturing and the assembly department; *total time allocated for producing these products*; and the *profit per product.*

The input data is provided in Figure 7.7, and this data must be collected before starting the analysis of a model.

Figure 7.7 Extended allocation model – input data, phase 1

Products	NT121	NT122	NT123	
Quantity of products to be produced				
Unique product characteristics				*Resources availability*
Raw material 1 from supplier 1 (units)	2	6	3	550
Raw material 2 from supplier 2 (units)	3	4	3	600
Raw material 3 from supplier 3 (units)	5	6	5	800
Time in manufacturing department (min)	20	20	10	2200
Time in the assembly department (min)	15	20	25	3500
Profit per product (£)	5	7	6	0

Scope. The scope of the analysis has been extended from the previous example and it captures the following:

- Identify the *total number of products* to be produced over a period of time.
- Maximize the *profit* to produce these three products over a period of three months and maximize the profit to produce six products over the period of the remaining three months.
- Maximize the *use of the resources* put in place for producing these products.
- Provide an *analysis* over a period of time.

Output data. The output data obtained at the end of the analysis is as follows:

- *Quantity to be produced per product per month.*
- *Quantity of raw material used in production per month.*
- *Leftover inventory of raw material per month.*
- *Total time spend in production per department per month.*
- *Resource utilization.*
- *Total profit per month.*

Variables. The variables in this case are the *total number of products to be manufactured*, for NT121, NT122 and NT123 each month. During phase 1 of analysis, a new review of each of these variables is to be provided.

- **Variable 1:** Total number of products to be product NT121.
- **Variable 2:** Total number of products to be product NT122.
- **Variable 3:** Total number of products to be product NT123.

Constraints. As in our previous example, the set of constraints are linked to supplier constraints and manufacturer constraints. There are three constraints for the raw materials used to produce these three products and two constraints for the manufacturing and assembly time required for production:

- **Raw material 1:** The total quantity of raw material 1 required to produce the three products is less than or equal to (<=) the quantity available in stock for raw material 1 (550 units).
- **Raw material 2:** The total quantity of raw material 2 required to produce the three products is less than or equal to (<=) the quantity available in stock for raw material 2 (in this case 600 units).
- **Raw material 3:** The total quantity of raw material 3 required to produce the three products is less than or equal to (<=) the quantity available in stock for raw material 3 (800 units).
- **Manufacturing time:** The total time required in the manufacturing department to produce the three products is less than or equal to (<=) the total allocated time (2,200 min).
- **Assembly time:** The total time required in the assembly department to produce the three products is less than or equal to (<=) the total allocated time (3,500 min).

The non-negative constraint is set as higher than or equal to (>=) value '0'. *Variable 1, Variable 2 and Variable 3 >= 0*

There is also the expectation that we will have 'integers' value for the three products produced. In other words, we will expect to work with full products.

- *Variable 1; Variable 2 and Variable 3 = integer*

Objective function. The objective function in this case is the *total profit* that is expected to be maximized.

In this case *max(Total Profit) = 5*Variable 1+7*Variable 2+6*Variable 3*

As previously, this model can be implemented and solved in Excel using the add-in software Solver.

The model template to be developed in this case can capture characteristics as indicated in Figure 7.8. The way in which these templates are developed within a spreadsheet depends on each individual preference; however, clear representation of variables, constraints and the objective function is key to clear communication and analysis. Figure 7.8 gives an example of how variables are clearly marked in cells C3 for product NT121, D3 for product NT122 and E3 for product NT123.

Figure 7.8 Extended allocation model – template, phase 1

	A	B	C	D	E	F	G	H	I
1	Analysis Phase 1- three products			Month 1					
2	Variables	Products	NT121	NT122	NT123				
3		*Quantity of products to be produced*							
4									
5		*Unique product characteristics*				*Used in production*		*Resources availability*	*Left over after production*
6	Constraints	Raw material 1 from supplier 1 (units)	2	6	3	0	<=	550	550
7		Raw material 2 from supplier 2 (units)	3	4	3	0	<=	600	600
8		Raw material 3 from supplier 3 (units)	5	6	5	0	<=	800	800
9		Time in manufacturing department (min)	20	20	10	0	<=	2200	2200
10		Time in the assembly department (min)	15	20	25	0	<=	3500	3500
11									
12	Objective	Profit per product (£)	5	7	6		Total profit	0	

The functions required to set the five constraints will be captured in cells F6 to F10 and these are:

- **Raw material 1 constraint:** Insert in cell F6 the following function = SUMPRODUCT(C6:E6,$C$3:$E$3)

- **Raw material 2 constraint:** Insert in cell F7 the function = SUMPRODUCT(C7:E7,$C$3:$E$3)
- **Raw material 3 constraint:** Insert in cell F8 the function = SUMPRODUCT(C8:E8,$C$3:$E$3)
- **Manufacturing time:** The required function in this case is = SUMPRODUCT(C9:E9,$C$3:$E$3)
- **Assembly time:** The function to be considered in this case is = SUMPRODUCT(C10:E10,$C$3:$E$3)

The objective function is to be inserted in cell H12 using the function: = SUMPRODUCT(C12:E12,C3:E3).

Figure 7.9 Extended allocation model – Solver, phase 1

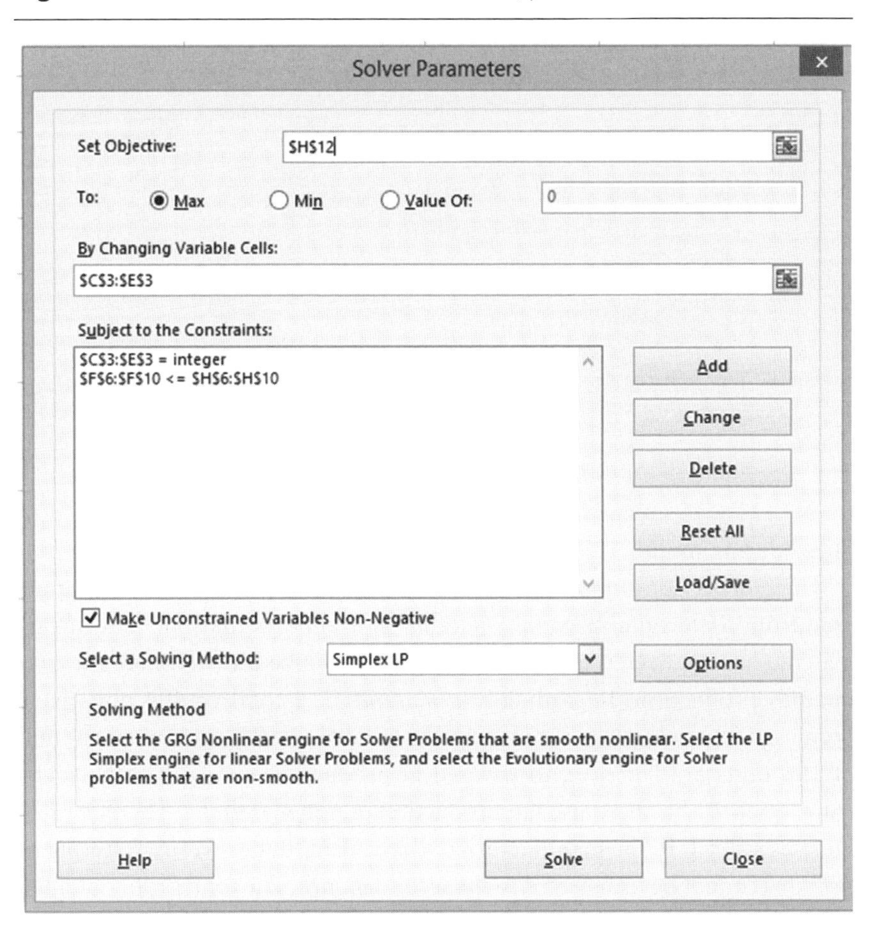

Having all these characteristics set up, the Excel add-in Solver software can be used to find the maximum profit. The settings required in Solver are as depicted in Figure 7.9. The value from cell H12 is considered for the *Set Objective* with the condition to be maximized. The *Changing Variable Cells* selects the values from cells C3 to E3. For the *constraints* the following are inserted: $F$6:$F$10 <= $H$6:$H$10 and for the variables we are looking to have these set as integer values. The *Mark Unconstrained Variables Non-Negative* option is selected and the solving method is selected as *Simplex LP*.

Solving this, the solution obtained for the settings during month 1 is presented in Figure 7.10.

Figure 7.10 Extended allocation model – solution for month 1, phase 1

	A	B	C	D	E	F	G	H	I
1	Analysis Phase 1- three products			Month 1					
2	Variables	Products	NT121	NT122	NT123				
3		*Quantity of products to be produced*	15	35	103				
4									
5		*Unique product characteristics*				*Used in production*		*Resources availability*	*Left over after production*
6	Constraints	Raw material 1 from supplier 1 (units)	2	6	3	549	<=	550	1
7		Raw material 2 from supplier 2 (units)	3	4	3	494	<=	600	106
8		Raw material 3 from supplier 3 (units)	5	6	5	800	<=	800	0
9		Time in manufacturing department (min)	20	20	10	2030	<=	2200	170
10		Time in the assembly department (min)	15	20	25	3500	<=	3500	0
11									
12	Objective	Profit per product (£)	5	7	6		Total profit	938	

The *quantity to be produced per product per month* represents the results for NT121 = 15 units; NT122 = 35 units and NT123 = 103 units to be produced during month 1.

The *quantity of raw material used in production per month* is 549 units used from 550 available of *raw material 1*. It appears that one unit of raw material has not been used in this case and this can be transferred to the next month's production. Regarding *raw material* 2, a total of 494 units have been used in production from the total available of 600 units, where a total of 106 units are left over, to be used in production during month 2. The entire quantity has been used in production of *raw material* 3. There are 170 minutes still left in the manufacturing department, but as we are referring to time this cannot be transferred to the second month. The *total profit* obtained during month 1 for producing these three products is £938.

At the beginning of setting up this analysis, the decision was made to place an order of the same quantity for each Raw material. As the production uses the entire stock from raw material 3, the optimization results for month 2 will remain the same regardless of the fact that there is leftover *raw material 1* and *raw material 2*. The decision in this case could be to increase the quantity supplied of raw material 3 to observe any changes in throughput, or continue with the same expectation of throughput, but reduce the quantity of order placed for the other raw materials supplied (*raw material 1* and *raw material 2*).

Setting up the analysis template for month 2, this could take the form as indicated in Figure 7.11. The same constraints are to be set within this month, still the functions used will need to follow the variables selected for this month.

The functions required to set the five constraints for month 2 will be captured in cells K6 to K10 and these are:

- **Raw material 1 constraint:** Insert in cell K6 the following function =SUMPRODUCT(C6:E6,$K$3:$M$3)
- **Raw material 2 constraint:** Insert in cell F7 the function =SUMPRODUCT(C7:E7,$K$3:$M$3)
- **Raw material 3 constraint:** Insert in cell F8 the function =SUMPRODUCT(C8:E8,$K$3:$M$3)
- **Manufacturing time:** The required function in this case is =SUMPRODUCT(C9:E9,$K$3:$M$3)
- **Assembly time:** The function to be considered in this case is =SUMPRODUCT(C10:E10,$K$3:$M$3)

The objective function for month 2 is to be inserted in cell M12 using the function: =SUMPRODUCT(C12:E12,K3:M3).

One other change that can be observed in this case is the quantity to be considered in cells M6 to M10 (see Figure 7.11). For the raw material constraints, the values to be considered are the order quantity placed + the quantity of materials left over from the previous month. Where no changes are to be considered to the order placed for the raw materials this will follow the functions: in cell M6 = H6 + I6, in cell M7 = H7 + I7 and in cell M8 = H8 + I8. Within cell M9, we see the same value for the allocated time as in month 1 and as in cell H9. As the reference in this case is related to time, this is not going to be a value that could be transferred to the following month.

Figure 7.11 Extended allocation model – template, month 2, phase 1

	A	B	C	D	E	F	G	H	I	J	K	L	M
1	**Analysis Phase 1- three products**			**Month 1**								**Month 2**	
2	**Variables**	Products	NT121	NT122	NT123						NT121	NT122	NT123
3		*Quantity of products to be produced*	15	35	103								
4													
5		*Unique product characteristics*				*Used in production*		*Resources availability*	*Left over after production*		*Used in production*		*Resources availability*
6	**Constraints**	Raw material 1 from supplier 1 (units)	2	6	3	549	<=	550	1		0	<=	605
7		Raw material 2 from supplier 2 (units)	3	4	3	494	<=	600	106		0	<=	706
8		Raw material 3 from supplier 3 (units)	5	6	5	800	<=	800	0		0	<=	880
9		Time in manufacturing department (min)	20	20	10	2030	<=	2200	170		0	<=	2200
10		Time in the assembly department (min)	15	20	25	3500	<=	3500	0		0	<=	3530
11													
12	**Objective**	Profit per product (£)	5	7	6		Total profit	938				Total profit	0

To further advance the analytics in this case, this model can operate with an assumed 10 per cent of safety stock. This is to be reflected in the resources available during month 2.

The condition considered in this case will be: if the leftover stock is less than or equal to the safety stock, then order the planned quantity + 10 per cent. If not, order the same quantity (see cells M6 to M8 in Figure 7.11).

For the condition of time, a 30 minute safety time is to be incorporated within the calculations. Therefore, the following condition can be considered here: if the value of time left over after production during month 1 is higher than or equal to 30 minutes, then no changes are to be considered to the time allocation for production in month 2. However, if this time is less than 30 minutes, an additional time of 30 minutes is to be considered for safety reasons (see cells M9 and M10 in Figure 7.11).

The same logic will be used to solve the model using Solver for months 2 and 3 respectively when a similar template for month 3 is to be developed. Figure 7.12 captures the results for the first three months of production. This figure also captures some of the performance measures as output data, such as the throughput for each product through the three months of production, the total profit over the three months, and inventory used and left over at the end of each month.

Following the same logic to modelling, three more products are to be included in the analysis for phase 2. The input data is displayed in Figure 7.13.

The mathematical model is then developed, following the same logic as mentioned before in phase 1. However, in this case we are going to work with six variables from *variable 1* being the number of products to be produced of NT121 to *variable 6* representing the number of products to be produced of NT133.

The same types of constraints are to be considered in this phase of analysis, where the objective function still remains to maximize profit. Therefore, in this case the mathematical model is represented as such:

Variables. The variables in this case are *the total number of products to be manufactured*, for NT121, NT122, NT123, NT131, NT132 and NT133 each month.

During phase 1 of analysis, a new review of each of these variables is to be provided.

- **Variable 1:** Total number of products to be products of NT121.
- **Variable 2:** Total number of products to be products of NT122.
- **Variable 3:** Total number of products to be products of NT123.

Figure 7.12 Extended allocation model – solution analysis, phase 1

	A	B	C	D	E	F	G	H	I	J	K	L	M	N	O	P	Q	R	S
1	Analysis Phase 1- three products		Month 1								Month 2					Month 3			
2	Variables	Products	NT121	NT122	NT123						NT121	NT122	NT123			NT121	NT122	NT123	
3		*Quantity of products to be produced*	15	35	103						14	50	92			16	50	91	
4																			
5		*Unique product characteristics*				*Used in production*		*Resources availability*	*Left over after production*		*Used in production*		*Resources availability*	*Left over after production*		*Used in production*		*Resources availability*	*Left over after production*
6	Constraints	Raw material 1 from supplier 1 (units)	2	6	3	549	<=	550	1		604	<=	605	1		605	<=	605	0
7		Raw material 2 from supplier 2 (units)	3	4	3	494	<=	600	106		518	<=	706	188		521	<=	788	267
8		Raw material 3 from supplier 3 (units)	5	6	5	800	<=	800	0		830	<=	880	50		835	<=	880	45
9		Time in manufacturing department (min)	20	20	10	2030	<=	2200	170		2200	<=	2200	0		2230	<=	2230	0
10		Time in the assembly department (min)	15	20	25	3500	<=	3500	0		3510	<=	3530	20		3515	<=	3530	15
11																			
12	Objective	Profit per product (£)	5	7	6		Total profit	938				Total profit	972				Total profit	976	

Total Profit per month

Profit £: 900, 920, 940, 960, 980

Month: 1, 2, 3

Throughput

0, 30, 60, 90, 120

NT121, NT122, NT123

Month 1, Month 2, Month 3

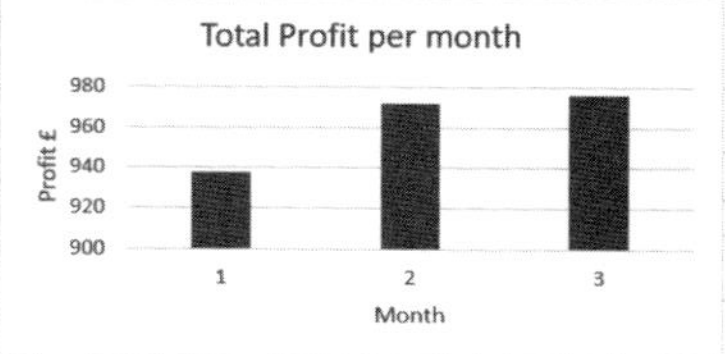

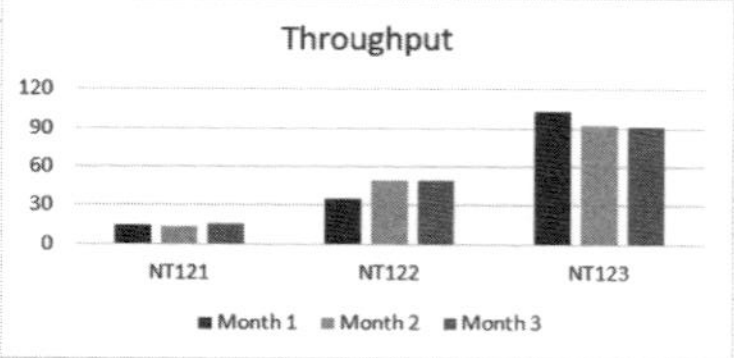

Figure 7.13 Extended allocation model – input data, phase 2

Analysis Phase 2- six products		Month 4						
Variables	Products	NT121	NT122	NT123	NT131	NT132	NT133	
	Quantity of products to be produced							
	Unique product characteristics							*Resources availability*
Constraints	Raw material 1 from supplier 1 (units)	2	6	3	3	7	6	1150
	Raw material 2 from supplier 2 (units)	3	4	3	5	5	4	1200
	Raw material 3 from supplier 3 (units)	5	6	5	3	6	2	1500
	Time in manufacturing department (min)	20	20	10	15	15	15	4700
	Time in the assembly department (min)	15	20	25	15	25	15	7000
Objective	Profit per product (£)	5	7	6	5	4	6	

During phase 2 of the analysis, the following variables will be considered:

- **Variable 4:** Total number of products to be products of NT131.
- **Variable 5:** Total number of products to be products of NT132.
- **Variable 6:** Total number of products to be products of NT133.

Constraints. As in our previous example, the set of constraints are linked to supplier constraints and manufacturer constraints. There are still three constraints for the raw material used to produce these six products and two constraints for the manufacturing and assembly time required for production.

- **Raw material 1:** The total quantity of raw material 1 required to produce the six products is less than or equal to (<=) the quantity available in stock for raw material 1 (1,150 units).
- **Raw material 2:** The total quantity of raw material 2 required to produce the six products is less than or equal to (<=) the quantity available in stock for raw material 2 (in this case 1,200 units).
- **Raw material 3:** The total quantity of raw material 3 required to produce the six products is less than or equal to (<=) the quantity available in stock for raw material 3 (1,500 units).
- **Manufacturing time:** The total time required in the manufacturing department to produce the six products is less than or equal to (<=) the total allocated time (4,700 min).
- **Assembly time:** The total time required in the assembly department to produce the six products is less than or equal to (<=) the total allocated time (7,000 min).

The non-negative constraint that refers to set the three variables to higher than or equal to (>=) value '0'.

Variable 1, Variable 2, Variable 3, Variable 4, Variable 5 and Variable 6 >= *0*

There is also the expectation that we will have 'integers' value for the three products produced. In other words, we will expect to work with full products.

Variable 1; Variable 2; Variable 3; Variable 4; Variable 5 and Variable 6 = integer

Objective function. The objective function in this case is the *total profit* that is expected to be maximized.

In this case *max(Total Profit) = 5*Variable 1 + 7*Variable 2 + 6*Variable 3 + 5*Variable 4 + 4*Variable 5 + 6*Variable 6*

Solving this example using Solver, the results are shown in Figure 7.14.

Figure 7.14 Extended allocation model – solution for month 4, phase 2

	A	B	C	D	E	F	G	H	I	J	K	L
1	**Analysis Phase 2- six products**		**Month 4**									
2	**Variables**	Products	NT121	NT122	NT123	NT131	NT132	NT133				
3		*Quantity of products to be produced*	68	1	176	48	0	56				
4		*Unique product characteristics*							*Used in production*		*Resources availability*	*Left over after production*
5	**Constraints**	Raw material 1 from supplier 1 (units)	2	6	3	3	7	6	1150	<=	1150	0
6		Raw material 2 from supplier 2 (units)	3	4	3	5	5	4	1200	<=	1200	0
7		Raw material 3 from supplier 3 (units)	5	6	5	3	6	2	1482	<=	1500	18
8		Time in manufacturing department (min)	20	20	10	15	15	15	4700	<=	4700	0
9		Time in the assembly department (min)	15	20	25	15	25	15	7000	<=	7000	0
10												
11	**Objective**	Profit per product (£)	5	7	6	5	4	6		Total profit	1979	

Figure 7.15 Extended allocation model – solution with specific demand, phase 2

	A	B	C	D	E	F	G	H	I	J	K	L
1	**Analysis Phase 2- six products**				**Month 4**							
2	**Variables**	Products	NT121	NT122	NT123	NT131	NT132	NT133				
3		*Quantity of products to be produced*	50	10	173	57	15	32				
4		*Unique product characteristics*							*Used in production*		*Resources availability*	*Left over after production*
5	**Constraints**	Raw material 1 from supplier 1 (units)	2	6	3	3	7	6	1147	<=	1150	3
6		Raw material 2 from supplier 2 (units)	3	4	3	5	5	4	1197	<=	1200	3
7		Raw material 3 from supplier 3 (units)	5	6	5	3	6	2	1500	<=	1500	0
8		Time in manufacturing department (min)	20	20	10	15	15	15	4490	<=	4700	210
9		Time in the assembly department (min)	15	20	25	15	25	15	6985	<=	7000	15
10												
11	**Objective**	Profit per product (£)	5	7	6	5	4	6		Total profit	1895	

What can be noted in this case is that the resources have been used in full, with the exception of raw material 3 where there are still 18 units left in stock.

This analysis can continue over a period of time, where the safety stock and time could be added in a logical way, as already described for phase 1 analysis.

One other element that can be introduced in this case is the demand from customers. If we now assume that demand for product NT122 is expected as a minimum of 10 units and the demand for product NT132 is expected as a minimum of 15 units, the results obtained after using Solver are as presented in Figure 7.15. These conditions are added in Solver as D3=12 and G3=15.

In this case the overall profit is lower, and not all resources have been utilized, however the customer demand is in line with the required specifications.

Covering model: Staff scheduling examples

Staff scheduling example 1: Resource allocation per week (weekly covering)

Following from the resource allocation example, let's assume that the time required in the manufacturing department requires particular attention. The manager is keen to correctly allocate staff time over a period of time (in this case a week) to meet the production requirements. Staff time is required not only for the two products detailed in the example above, but also for other products and activities required within a department. Therefore, the example developed below shows how resources could be allocated to cover the requirements for an entire week during the production.

In this case it is assumed that the minimum number of hours required in a day is known and set in the production plan. Considering that each person is expected to work the same number of hours a day, the total number of hours required is split and represented as the total number of operators/workers required in a particular day. The second condition known is that each person is expected to work five consecutive days and then take two days off. To avoid interrupting the production, Saturdays and Sundays are also captured in the production planning. The input data is presented in Figure 7.16.

Figure 7.16 Covering model – staff scheduling example 1 – input data

Allocation per day	Working patten per day							
	Monday	Tuesday	Wednesday	Thursday	Friday	Saturday	Sunday	Minimum number of operators required/day
Monday	1	0	0	1	1	1	1	8
Tuesday	1	1	0	0	1	1	1	9
Wednesday	1	1	1	0	0	1	1	7
Thursday	1	1	1	1	0	0	1	11
Friday	1	1	1	1	1	0	0	9
Saturday	0	1	1	1	1	1	0	4
Sunday	0	0	1	1	1	1	1	4

The vertical columns indicate the days worked, so if an operator starts work on a Monday, they will work on Tuesday, Wednesday, Thursday and Friday, and will take Saturday and Sunday off. If an operator starts work on a Tuesday, they will work for five consecutive days until Saturday and will take Sunday and Monday off. This pattern will continue this way for the entire week, as indicated in Figure 7.16.

The horizontal rows in this figure will represent who works on a Monday. So, we will have operators starting their work on a Monday, as well as operators starting their work on a Thursday, Friday, Saturday and Sunday. Those operators starting work on a Tuesday and a Wednesday will not be working on Monday.

Input data. The required input data in this case will be the *total number of hours required for each particular day*, or the *total number of operators required on each day of the week*.

Goal. The goal in this case is to find the *optimum number of operators required in a week* following the conditions mentioned.

Output data. The expected output data will be:

- *Maximum number of operators who work in a particular week.*
- *Total number of operators who start their work on a particular day.*
- *Total number of operators per week.*

To solve this example using the linear programming approach we will need to identify the variables, constraints and the objective function.

Variables. The variables in this case are:

$$x_1, x_2, x_3, x_4, x_5, x_6, x_7$$

Where x1 represents the number of operators starting their work on a Monday; x2 is the number of operators starting their work on a Tuesday; x3 number of operators starting their work on a Wednesday; and so on, finishing with x7 representing the number of operators starting their work on a Sunday.

Constraints. The following constraints could be defined in this case as follows:

Operators working on Mondays will be those that start their first working day on a Monday, Thursday, Friday, Saturday and Sunday: $x_1 + x_2 + x_3 + x_4 + x_5 + x_6 + x_7 \geq 8$

Operators working on Tuesdays: $x_2 + x_5 + x_6 + x_7 + x_1 \geq 9$

Operators working on Wednesdays: $x_3 + x_6 + x_7 + x_1 + x_2 \geq 7$

Operators working on Thursdays: $x_4 + x_7 + x_1 + x_2 + x_3 \geq 11$

Operators working on Fridays: $x_5 + x_1 + x_2 + x_3 + x_4 \geq 9$

Operators working on Saturdays: $x_6 + x_2 + x_3 + x_4 + x_5 \geq 4$

Operators working on Sundays: $x_7 + x_3 + x_4 + x_5 + x_6 \geq 4$

To the list of constraints presented above we will also add the following two:

$$x_1, x_2, x_3, x_4, x_5, x_6, x_7 \geq 0 \textit{ and}$$
$$x_1, x_2, x_3, x_4, x_5, x_6, x_7 = \textit{integers values}$$

To be able to solve this problem using Excel, the template shown in Figure 7.17 can be set up to capture all the required information.

Objective function. The objective function is to find the minimum number of operators who staff this department per week, when all the constraints are met.

In Excel the *variables* will be set under the cells: C3, D3, E3, F3, G3, H3 and I3.

The *constraints* can be set as follows:

Operators working on Mondays: J8 = SUMPRODUCT(C8:I8,$C$3:$I$3). In Solver, we will set J8 >= L8.

Operators working on Tuesdays: J9 = SUMPRODUCT(C9:I9,$C$3:$I$3). In Solver, we will set J9 >= L9.

Operators working on Wednesdays: J10 = SUMPRODUCT(C10:I10,$C$3:$I$3). In Solver, we will set J10 >= L10.

Operators working on Thursdays: J11 = SUMPRODUCT(C11:I11,$C$3:$I$3). In Solver, we will set J11 >= L11.

Operators working on Fridays: J12 = SUMPRODUCT(C12:I12,$C$3:$I$3). In Solver, we will set J12 >= L12.

Figure 7.17 Covering model – staff scheduling example 1 – Excel template

	A	B	C	D	E	F	G	H	I	J	K	L
1		***Variables***										
2			Monday	Tuesday	Wednesday	Thursday	Friday	Saturday	Sunday			
3												
4		*used as notations only	x_1	x_2	x_3	x_4	x_5	x_6	x_7			
5												
6		***Constraints***	*Working patten per day*									
7			Monday	Tuesday	Wednesday	Thursday	Friday	Saturday	Sunday	*Acctual operators per day*		*Minimum number of operators required/day*
8	*Allocation per day*	Monday	1	0	0	1	1	1	1	0	>=	8
9		Tuesday	1	1	0	0	1	1	1	0	>=	9
10		Wednesday	1	1	1	0	0	1	1	0	>=	7
11		Thursday	1	1	1	1	0	0	1	0	>=	11
12		Friday	1	1	1	1	1	0	0	0	>=	9
13		Saturday	0	1	1	1	1	1	0	0	>=	4
14		Sunday	0	0	1	1	1	1	1	0	>=	4
15												
16		***Objectives***	0									

Operators working on Saturdays: J13 = SUMPRODUCT(C13:I13,$C$3:$I$3). In Solver, we will set J13 >= L13.

Operators working on Sundays: J14 = SUMPRODUCT(C14:I14,$C$3:$I$3). In Solver, we will set J14 >= L14.

The other two constraints to be added in Solver are: C3:I3 = integer, and to set that they >= 0, selecting the condition *Make Unconstrained Variables Non-Negative* will be sufficient.

The objective function is set in cell C16 using = SUM(C3:I3).

The Solver will appear as presented in Figure 7.18.

Figure 7.18 Covering model – staff scheduling example 1 – Solver

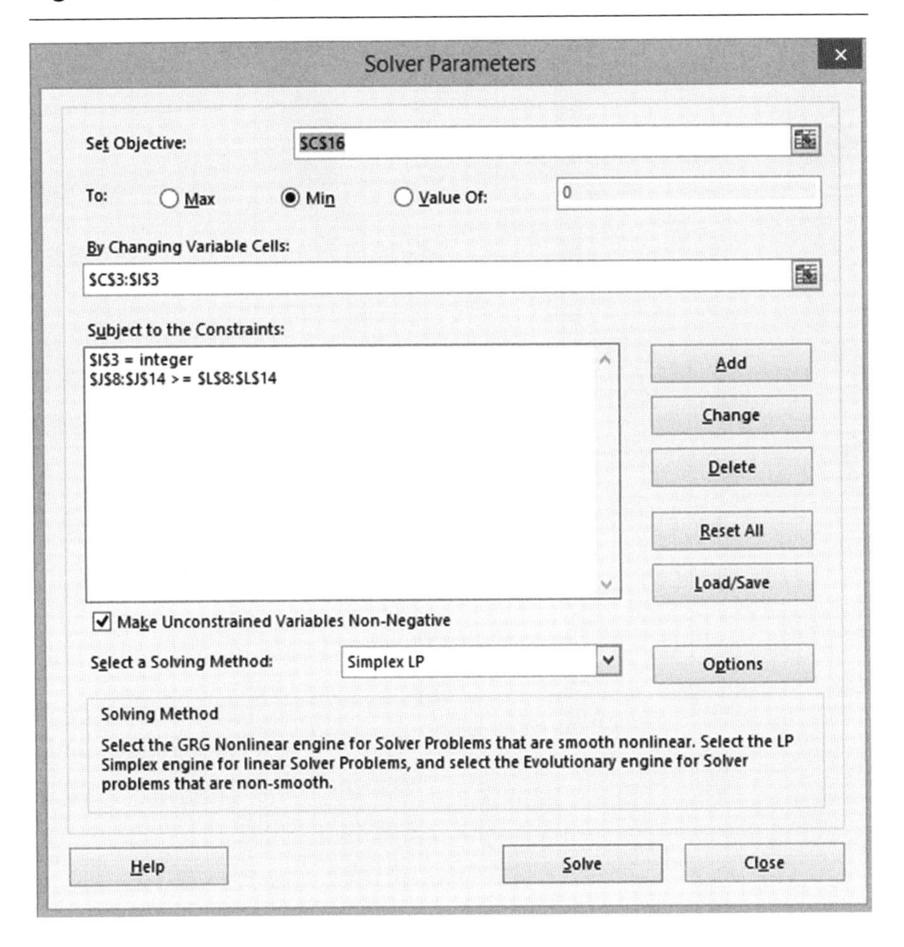

The final results obtained in this case are as indicated in Figure 7.19.

Figure 7.19 Covering model – staff scheduling example 1 – solution

	A	B	C	D	E	F	G	H	I	J	K	L
1		*Variables*										
2			Monday	Tuesday	Wednesday	Thursday	Friday	Saturday	Sunday			
3			4	3	0	2	0	0	2			
4		*used as notations only	x_1	x_2	x_3	x_4	x_5	x_6	x_7			
5												
6		*Constraints*	*Working patten per day*									
7			Monday	Tuesday	Wednesday	Thursday	Friday	Saturday	Sunday	*Acctual operators per day*		*Minimum number of operators required/day*
8	*Allocation per day*	Monday	1	0	0	1	1	1	1	8	>=	8
9		Tuesday	1	1	0	0	1	1	1	9	>=	9
10		Wednesday	1	1	1	0	0	1	1	9	>=	7
11		Thursday	1	1	1	1	0	0	1	11	>=	11
12		Friday	1	1	1	1	1	0	0	9	>=	9
13		Saturday	0	1	1	1	1	1	0	5	>=	4
14		Sunday	0	0	1	1	1	1	1	4	>=	4
15												
16		*Objectives*	11									

The solution obtained gives information on the total number of operators required for a particular week. In this case, the value obtained is 11 operators required per week. Within cells J8 to J14, the number of operators working on each day in a week is also indicated. To meet the overall constraints, nine operators are working on Wednesday as opposed to seven originally planned, and five operators are working on Saturday as opposed to four originally planned. For all other days, the minimum number of operators required per day matches the optimization solution obtained.

Our model also indicates the number of operators starting their working day on each particular day. Therefore, four operators will start their working days on a Monday, three on a Tuesday, two on a Thursday, one and two on a Sunday. Based on this solution, no operators are starting their working day on a Friday or a Saturday.

It is relevant to note that this is one solution Solver can indicate; however, Solver may result in a different combination of allocated operators per day, still having the same total value for the number of operators per week.

Observations. A number of scenarios could be studied using the covering model where changes to the input data could be considered.

For example, modifications to the model could consider changes in the working pattern, such as operators are expected to work a reduced number of days before they take a break (such as work three days and take two days off, followed by another three days working and two days off, and so on).

One other modification to the model could be of a numerical nature, where the number of operators required per day is changing. For example, on Monday 20 operators are now required as opposed to eight in the original setting, and 15 operators are required on Tuesday instead of the original nine, and so on. The settings of the model (from the set inequalities, to the formulas used in Excel, to the settings considered in Solver) will remain exactly the same. However, the results will have a set of different values.

A similar example to the one presented above can be found in Albright and Winston (2011), the name worker scheduling model, where an example of postal employee scheduling is presented.

Staff scheduling example 2: Resource allocation per day (daily covering)

The daily covering example can be explained in the context of the same example discussed above, or for any other example that required staff to cover activities with particular emphasis for a particular day. Call centres may use this approach as they operate 24 hours per day, where it is important to have the right coverage per day. This example can also be applied to operating a warehouse, or in the case of a production line that is open 24 hours a day. These examples can also be seen in ports, airports, and train stations. There are also a number of situations outside the field of logistics and supply chain where these examples can be used.

To define the problem, let's assume that for a particular logistics department that offers services 24 hours, the data defined in Figure 7.20 represents the operations requirements. A working day (24 hours) is split in six shifts, each shift representing four hours. Staff is expected to work two consecutive shifts and take the rest of the day off.

Figure 7.20 Covering model – staff scheduling example 2 – input data

		Working patten per shift						
		7am-11am	11am-15pm	15pm-19pm	19pm-23pm	23pm-3am	3am-7am	*Minimum number of operators required/shift*
Allocation per shift	7am-11am	1	0	0	0	0	1	15
	11am-15pm	1	1	0	0	0	0	20
	15pm-19pm	0	1	1	0	0	0	18
	19pm-23pm	0	0	1	1	0	0	10
	23pm-3am	0	0	0	1	1	0	5
	3am-7am	0	0	0	0	1	1	3

As in the previous example, indicate who works on a particular shift per day. So, from 7 am to 11 am the operators who started their working shift at 7 am are considered in this shift, and the operators who started their work at 3 am are also working during the shift at 7 am to ensure they complete their 8 hours' required work per day. The same goes for the following shift from 11 am to 3 pm. The operators working during this shift are those who started their working day during the previous shift at 7 am, and the operators who started their shift at 11 am. This pattern will continue until all the allocated working shifts have been identified. Therefore, during the shift from 3 am to 7 am, the operators starting their shift at 3 pm and the operators starting their shift at 3 am are working during this shift.

The vertical columns indicate the working pattern for operators who started their shift at a particular time. For example, those who started their work at 7 am work first shift from 7 am to 11 am, the second shift from 11 am to 3 pm and so on.

Input data. The required input data is the total number of operators required on each shift per day.

Goal. The goal is to find *the optimum number of operators required in a day* following the conditions mentioned.

Output data. The expected output data will be:

- *Maximum number of operators that work in a particular shift.*
- *Total number of operators who work on any particular shift per day.*
- *Total number of operators per day.*

To solve this example using the linear programming approach we will need to identify the variables, constraints and the objective function.

Variables. The variables in this case for the six identified shifts per day are:

$$x_1, x_2, x_3, x_4, x_5, x_6$$

Where x1 represents the number of operators starting their work from 7 am to 11 am; x2 is the number of operators starting their shift from 11 am to 3 pm; x3 number of operators starting their work from 3 pm to 7 pm; x4 the number of operators starting their shift from 7 pm to 11 pm; x5 the number of operators starting their shift from 11 pm to 3 am; and x6 the number of operators starting their shift from 3 am to 7 am.

Constraints. The following constraints are:

Working on shift 7 am–11 am: $x_1 + x_6 \geq 15$

Working on shift 11 am–3 pm: $x_1 + x_2 \geq 20$

Working on shift 3 pm–7 pm: $x_2 + x_3 \geq 18$

Working on shift 7 pm–11 pm: $x_3 + x_4 \geq 10$

Working on shift 11 pm–3 am: $x_4 + x_5 \geq 5$

Working on shift 3 am–7 am: $x_5 + x_6 \geq 3$

To the list of constraints presented above we will also add the following two:

$$x_1, x_2, x_3, x_4, x_5, x_6 \geq 0 \textit{ and}$$

$$x_1, x_2, x_3, x_4, x_5, x_6 = \textit{integers values}$$

To be able to solve this problem using Excel, the template in Figure 7.21 can be set up to capture all the required information.

Figure 7.21 Covering model – staff scheduling example 2 – Excel template

	A	B	C	D	E	F	G	H	I	J	K
1		***Variables***									
2		Shifts per day	7am-11am	11am-15pm	15pm-19pm	19pm-23pm	23pm-3am	3am-7am			
3											
4		*used as notations only*	x_1	x_2	x_3	x_4	x_5	x_6			
5											
6		***Constraints***	*Working patten per shift*								
7			7am-11am	11am-15pm	15pm-19pm	19pm-23pm	23pm-3am	3am-7am	*Acctual operators per shift*		*Minimum number of operators required/shift*
8	*Allocation per shift*	7am-11am	1	0	0	0	0	1	0	>=	15
9		11am-15pm	1	1	0	0	0	0	0	>=	20
10		15pm-19pm	0	1	1	0	0	0	0	>=	18
11		19pm-23pm	0	0	1	1	0	0	0	>=	10
12		23pm-3am	0	0	0	1	1	0	0	>=	5
13		3am-7am	0	0	0	0	1	1	0	>=	3
14											
15		***Objectives***	0								

Objective function. The objective function is to find the minimum number of operators that staff this department per day, when all the constraints are met.

In Excel the *variables* will be set under the cells: C3, D3, E3, F3, G3 and H3.

The *constraints* can be set as follows:

Working on shift 7 am–11am: I8 = SUMPRODUCT(C8:H8,$C$3:$H$3).
In Solver, we will set I8 >= K8.

Working on shift 11 am–3 pm: I9 = SUMPRODUCT(C9:H9,$C$3:$H$3).
In Solver, we will set I9 >= K9.

Working on shift 3 pm–7 pm: I10 = SUMPRODUCT(C10:H10,$C$3:$H$3).
In Solver, we will set I10 >= K10.

Working on shift 7 pm–11 pm: I11 = SUMPRODUCT(C11:H11,$C$3:$H$3).
In Solver, we will set I11 >= K11.

Working on shift 11 pm–3 am: I12 = SUMPRODUCT(C12:H12,$C$3:$H$3).
In Solver, we will set I12 >= K12.

Working on shift 3 am–7 am: I13 = SUMPRODUCT(C13:H13,$C$3:$H$3).
In Solver, we will set I13 >= K13.

The other two constraints to be added in Solver are: C3:H3 = integer, and to set that they >= 0, selecting the condition *Make Unconstrained Variables Non-Negative* will be sufficient.

The *objective function* is set in cell C15 using = SUM(C3:H3).

The Solver will appear as presented in Figure 7.22.

After running Solver, the final results can be observed in Figure 7.23.

The solution obtained gives information on the total number of operations required on a particular day. In this case the value obtained is 38. Within the cells I8 to I13, the number of operators working on each shift is also indicated. To meet the overall constraints, the minimum number of operators required on the shift between 7 pm and 11 pm is 10, but the model returns an answer as 13. This still meets the model constraints; however, it is relevant to know that during this particular shift there are three operators that may be allocated to other tasks.

Our model also indicated that 15 operators will start their working days during the first shift from 7 am, five will start their working day at 11 am, 13 will start at 15 pm and five will start at 23 pm.

Observations. As observed in Example 1, a number of changing scenarios could be studied in this case where, again, a number of changes could be considered to the input data.

One modification to the model could consider changes in the working pattern of the model, such as the total number of operators expected to

Figure 7.22 Covering model – staff scheduling example 2 – Solver

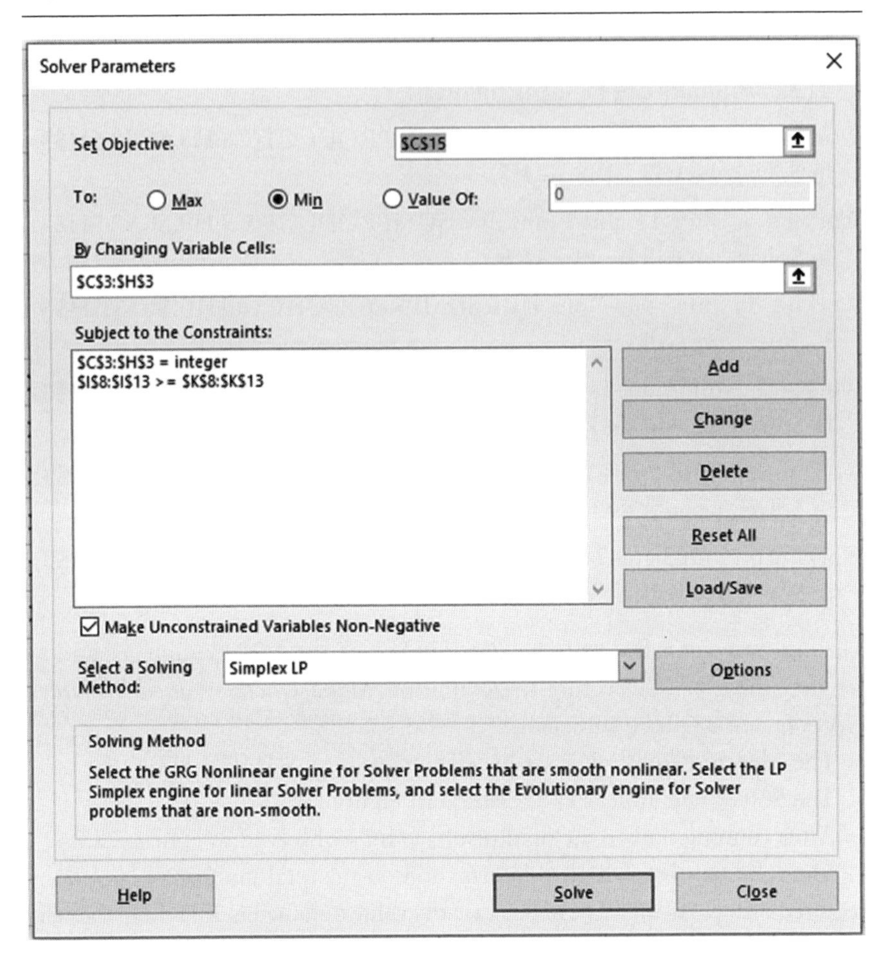

work on a particular shift. The mathematical model and the Solver settings of the model will remain exactly the same; however, it is expected that the results will have a different set of numerical values.

The model could also be set to indicate that no operators are required during the night shifts, or that some operators only work one shift, and so on.

Two examples of the covering model have been detailed here, but this model can be applied to many situations and different types of changes. Similar examples where the covering model has been presented can be further explored in Albright and Winston (2011) for weekly cover or Baker (2016) for daily scheduling and other variations.

Figure 7.23 Covering model – staff scheduling example 2 – solution

	A	B	C	D	E	F	G	H	I	J	K
1		*Variables*									
2		Shifts per day	7am-11am	11am-15pm	15pm-19pm	19pm-23pm	23pm-3am	3am-7am			
3			15	5	13	0	5	0			
4		*used as notations only	x_1	x_2	x_3	x_4	x_5	x_6			
5											
6		*Constraints*	*Working patten per shift*								
7			7am-11am	11am-15pm	15pm-19pm	19pm-23pm	23pm-3am	3am-7am	*Acctual operators per shift*		*Minimum number of operators required/shift*
8	*Allocation per shift*	7am-11am	1	0	0	0	0	1	15	>=	15
9		11am-15pm	1	1	0	0	0	0	20	>=	20
10		15pm-19pm	0	1	1	0	0	0	18	>=	18
11		19pm-23pm	0	0	1	1	0	0	13	>=	10
12		23pm-3am	0	0	0	1	1	0	5	>=	5
13		3am-7am	0	0	0	0	1	1	5	>=	3
14											
15		*Objectives*	38								

Assignment model

The assignment model is another linear programming model that is used in the field of supply chain. For example, consider a large manufacturing organization with manufacturing plants in many locations that is currently producing a range of different categories of products at each plant. To further optimize the operation, it is looking to explore whether producing just one category of products at a dedicated plant will optimize overall the cost of production. The cost to produce a particular range of products from a particular category at each plant has been estimated and provided in Figure 7.24. To explore this type of change, the analysis at this organization will aim to employ the assignment model.

Figure 7.24 Assignment model – input data

	Annual Cost (millions)	Product categoy (PC)						
		PC1	**PC2**	**PC3**	**PC4**	**PC5**	**PC6**	**PC7**
Manufacturing location (ML)	**ML1**	120	45	89	65	110	72	57
	ML2	98	67	85	63	112	76	59
	ML3	98	48	74	61	98	68	52
	ML4	110	58	72	54	97	71	56
	ML5	87	62	69	54	98	59	58
	ML6	95	79	68	56	109	57	61
	ML7	58	49	85	64	89	79	68

Input data. The model's input data required by analysis is the *cost to produce a particular category of products at a particular plant.* This value can be an estimated value based on producing any other similar type of products at a particular plant.

Goal. The goal is to *minimize the overall cost of production* when only one category of products is produced at a particular plant.

Output data. Therefore, the model's output data will be:

- *Total cost of production.*
- *The location assigned to produce a category of products.*

The model in this case can be defined as follows.

Variables. The possible location at which a category of products can be produced. In this particular example 49 variables are identified.

Constraints. There are two sets of constraints: one for the manufacturing location and one for the product category. As there are seven manufacturing locations, we can say that we are working with seven constrains linked to the manufacturing location. Similarly, as there are seven different product categories, we can say we are working with seven constraints linked to the product categories. Therefore, in total we have 14 constraints. As the aim is to assign only one product category per plant, then we have:

- *Constraint set 1:* A product category to be assigned to minimum one manufacturing location – the total sum of product category (PC1, PC2, …, PC7) >= 1.
- *Constraint set 2:* A plant has only one product category assigned to them – the total number of PCs allocated to one manufacturing location (ML1, ML2, … ML7) <= 1.

Objective function. The total cost of producing a category of products at only one location is minimized.

In Excel the template shown in Figure 7.25 can be developed.

Figure 7.25 Assignment model – Excel template

	A	B	C	D	E	F	G	H	I	J	K	L
1	**Assignment model**											
2												
3			Product categoy (PC)									
4		Annual Cost (millions)	**PC1**	**PC2**	**PC3**	**PC4**	**PC5**	**PC6**	**PC7**			
5	Manufacturing location (ML)	**ML1**	120	45	89	65	110	72	57			
6		**ML2**	98	67	85	63	112	76	59			
7		**ML3**	98	48	74	61	98	68	52			
8		**ML4**	110	58	72	54	97	71	56			
9		**ML5**	87	62	69	54	98	59	58			
10		**ML6**	95	79	68	56	109	57	61			
11		**ML7**	58	49	85	64	89	79	68			
12												
13	**Assignment decisions (variable and constraints)**											
14			**PC1**	**PC2**	**PC3**	**PC4**	**PC5**	**PC6**	**PC7**			Assigned
15		**ML1**								0	<=	1
16		**ML2**								0	<=	1
17		**ML3**								0	<=	1
18		**ML4**								0	<=	1
19		**ML5**								0	<=	1
20		**ML6**								0	<=	1
21		**ML7**								0	<=	1
22			0	0	0	0	0	0	0			
23			>=	>=	>=	>=	>=	>=	>=			
24		Assigned	1	1	1	1	1	1	1			
25												
26	**Objective function**											
27		Total cost	0									

The *variables* are set in cells C15 to I21.

Constraint set 1: are all set in cells C22 to I22 as:

C22 =SUM(C15:C21) >= C24

...

I22 =SUM(I15:I21) >= I24

Constraint set 2 are all set in cells J15 to J21:

J15 =SUM(C15:I15) <= 1

...

J21 =SUM(C21:I21) <= 1

The *objective function* is set in cell C27 with the following formula: =SUMPRODUCT(C5:I11,C15:I21)

All of these details are then inserted in Solver, and Solver will have the representation shown in Figure 7.26).

Figure 7.26 Assignment model – Solver

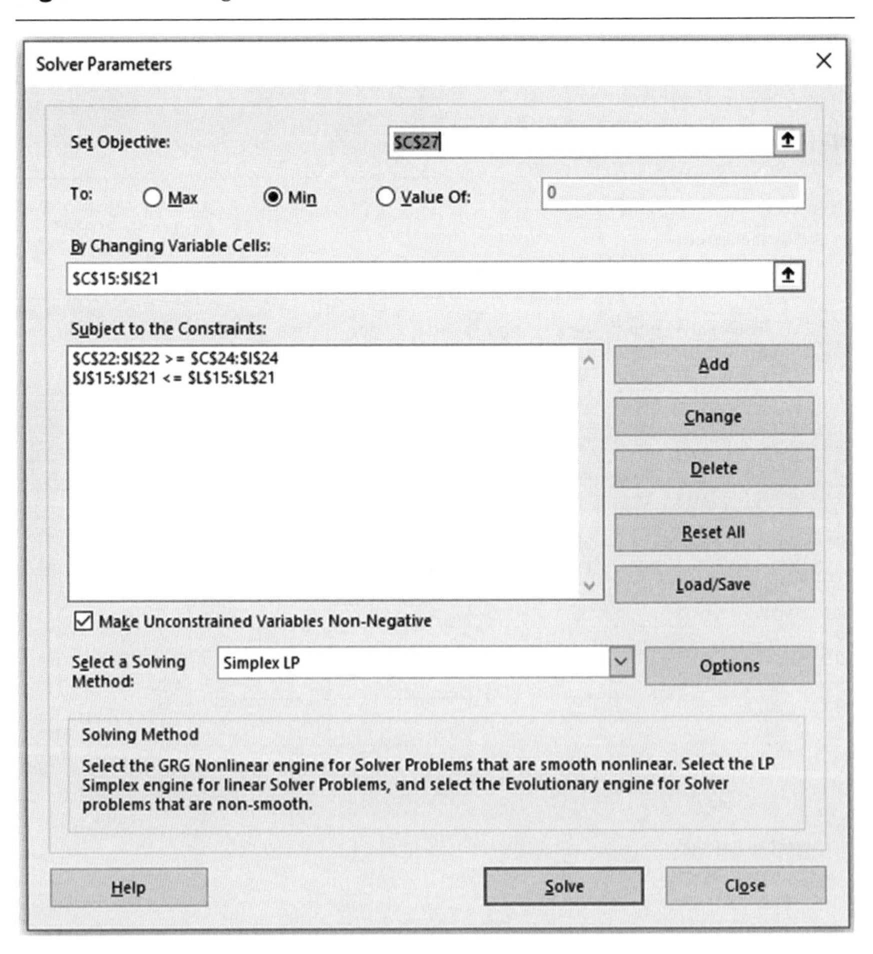

Following Solver, the solution is depicted in Figure 7.27. It can be noted that only one category of products has been allocated to a manufacturing plant location.

Figure 7.27 Assignment model – solution

	A	B	C	D	E	F	G	H	I	J	K	L
1	**Assignment model**											
2												
3			Product categoy (PC)									
4		Annual Cost (millions)	**PC1**	**PC2**	**PC3**	**PC4**	**PC5**	**PC6**	**PC7**			
5	Manufacturing location (ML)	**ML1**	120	45	89	65	110	72	57			
6		**ML2**	98	67	85	63	112	76	59			
7		**ML3**	98	48	74	61	98	68	52			
8		**ML4**	110	58	72	54	97	71	56			
9		**ML5**	87	62	69	54	98	59	58			
10		**ML6**	95	79	68	56	109	57	61			
11		**ML7**	58	49	85	64	89	79	68			
12												
13	**Assignment decisions (variable and constraints)**											
14			**PC1**	**PC2**	**PC3**	**PC4**	**PC5**	**PC6**	**PC7**			Assigned
15		**ML1**	0	1	0	0	0	0	0	1	<=	1
16		**ML2**	0	0	0	0	0	0	1	1	<=	1
17		**ML3**	0	0	0	0	1	0	0	1	<=	1
18		**ML4**	0	0	0	1	0	0	0	1	<=	1
19		**ML5**	0	0	1	0	0	0	0	1	<=	1
20		**ML6**	0	0	0	0	0	1	0	1	<=	1
21		**ML7**	1	0	0	0	0	0	0	1	<=	1
22			1	1	1	1	1	1	1			
23			>=	>=	>=	>=	>=	>=	>=			
24		Assigned	1	1	1	1	1	1	1			
25												
26	**Objective function**											
27		Total cost	440									

Observations. This is a model that can be used to assess the assignment (in this case) of products to particular manufacturing locations, or it can be used to assign services to particular sites, and so on. Different changes can be considered to this model, such as more product categories, and the number of manufacturing plant locations can be increased or decreased.

Other examples of the assignment model can be found in Albright and Winston (2011), where they present models on assigning machines to jobs to ensure the total completion time is being minimized, an example that can be applied to manufacturing job-shop operations. One other assignment example presented in the next section is assigning school buses to routes, which is also relevant to the field of logistics.

Direct transportation model

In the supply chain field, this model is applied in many situations that require the planning of transporting products from suppliers to manufacturers, or from the manufacturers to distribution centres, or from distribution centres to retailers, and so on. In other words, this model is used when products are transported from locations called sources or origins, to destinations.

The model assumes that there are capacity limitations at the source from where products are to be transported to the final destination. The cost to transport products from the source to the final destination forms part of the input data for this model and the customer demand values will also form part of the model's expected input data.

To further detail the applicability of such a model in the field of supply chain, the following example is developed.

Direct transportation example.

Let's assumes that the NTmanufacturer assembles products at three factories named FactoryNT1, FactoryNT2 and FactoryNT3. Products produced at these factories are then distributed to five distribution centres around the country, named DC1, DC2, DC3, DC4 and DC5, as presented in Figure 7.28.

Figure 7.28 Direct transportation example

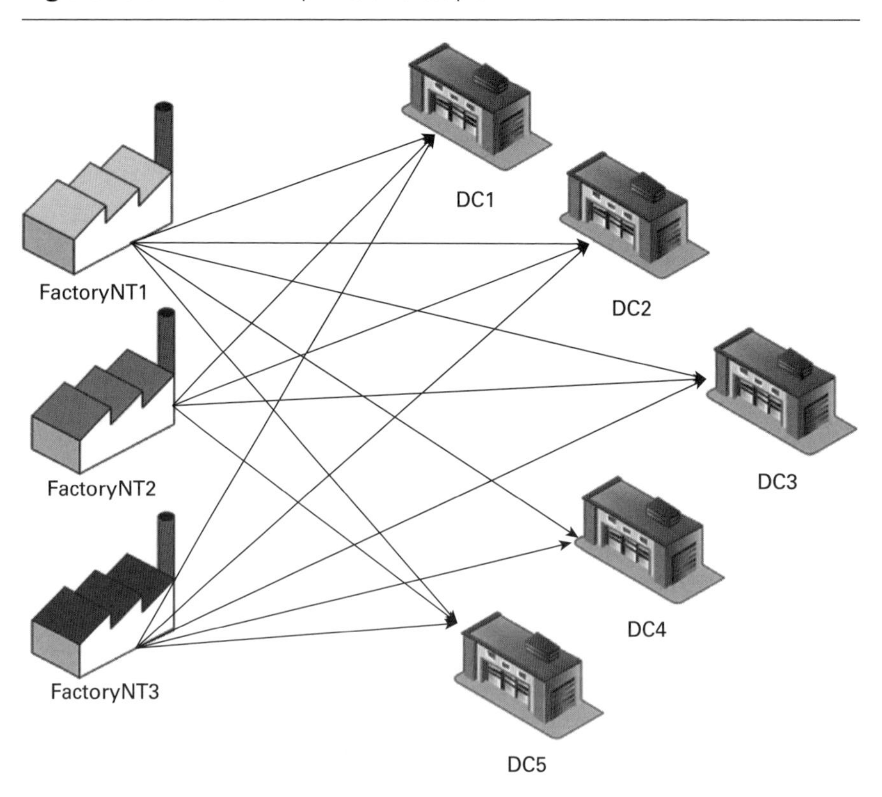

For this type of model, the analysis will only be focused on this section of the supply chain system.

Issues. The set of issues that have been identified by the NTmanufacturer and can be addressed by using this type of model are:

- The NTmanufacturer has observed a high transportation cost and aims to reduce this.
- It requires a clear plan for transporting the products to distribution centres to avoid committing to transporting volume above factory capacity.
- It wants to maintain 100 per cent customer service and aims to deliver on time in full.

Input data. For this model the following input data (see Figure 7.29) must be collected:

- The *transportation cost per product* from sources to destination.
- *Capacity at source*, in other words the maximum capacity to assemble products at each factory.
- *Demand at destination*, therefore the products demanded by each distribution centre based on their customers' demand.

Figure 7.29 Direct transportation example – input data

	A	B	C	D	E	F	G	H
1	*Input Data*							
2		**Cost/unit**	**DC1**	**DC2**	**DC3**	**DC4**	**DC5**	*Plant Capacity*
3		FactoryNT1	13	21	25	20	14	450
4		FactoryNT2	10	24	11	26	27	500
5		FactoryNT3	17	18	21	12	13	600
6		*Customer Demand*	350	245	400	200	150	

Goal. The goal in this case is:

- Identify the minimum transportation cost when delivering products from the factories to the distribution centres, while not exceeding the factory capacity.
- Meet the customer demand in full.
- Provide an optimum delivery plan for transporting products.

Output data. The expected output data at the end of the analysis is as follows:

- *Quantity to be delivered from each factory to each distribution centre.*
- *Total quantity delivered from each factory and the leftover after a delivery.*
- *Total transportation cost value.*
- *Total number of products delivered to each distribution centre.*

The mathematical model in this case is as follows.

Variables. The following variables are identified:

The number of products transported from sources to destinations.

In our case we have three sources and five destinations, therefore 15 variables.

Constraints. There are two sets of constraints that characterizes the direct transportation example. These are:

Constraints Set 1. The number of products transported from the sources to the destinations needs to be less than or equal to the number of products available at the source.

In our case there are three individual constraints that will follow this:

Total number of products transported from FactoryNT1 to the five distribution centres <= 450 (the total available at this factory)

Similarly, we will have for the other two factories:

Total number of products transported from FactoryNT2 to the five distribution centres <= 500

Total number of products transported from FactoryNT3 to the five distribution centres <= 600

Constraints Set 2. The total number of products transported to the destinations is higher than or equal to the demand at destination.

As we have five destinations, five distribution centres we transport products to, we will have five constraints in this case as follow:

Total number of products transported to DC1 >= Demand at DC1

Total number of products transported to DC2 >= Demand at DC2

Total number of products transported to DC3 >= Demand at DC3

Total number of products transported to DC4 >= Demand at DC4

Total number of products transported to DC5 >= Demand at DC5

On top of these eight constraints, two more are to be added: to set the variables >= 0 and to have integer values.

Objective function. The objective function in this case will be: *minimize the total transportation cost.*

Setting this example in Solver, the template developed in Figure 7.30 can be used.

Figure 7.30 Direct transportation example – Excel template

	A	B	C	D	E	F	G	H	I	J
1	Input Data									
2		**Cost/unit**	**DC1**	**DC2**	**DC3**	**DC4**	**DC5**			
3		FactoryNT1	13	21	25	20	14			
4		FactoryNT2	10	24	11	26	27			
5		FactoryNT3	17	18	21	12	13			
6										
7	Distribution plan									
8		**Transportation Plan**	**DC1**	**DC2**	**DC3**	**DC4**	**DC5**	*Total volume transported*		*Plant Capacity*
9		FactoryNT1							<=	450
10		FactoryNT2							<=	500
11		FactoryNT3							<=	600
12		*Total received*								
13			>=	>=	>=	>=	>=			
14		*Customer Demand*	350	245	400	200	150			
15										
16		*Total Transportation cost*								

The *variables* are represented in cells C9, C10, C11 for the products transported to DC1 from the three factories, and so on, products transported to the DC5 are listed under G9, G10 and G11.

For the first set of constraints, named as plant capacity constraints, the following formulas are to be inserted as: in cell H9 = SUM(C9:G9); in cell H10 = SUM(C10:G10); and in cell H11 = SUM(C11:G11).

The *set 1 constraints* will be:

H9 <= J9

H10 <= J10

H11 <= J11

For the second set of constraints, we will need to set up the following formulas in Excel: in cell C12 = SUM(C9:C11); D12 = SUM(D9:D11) and so on until cell G12 = SUM(G9:G11).

The *set 2 constraints* are:

C12 >= C14

D12 >= D14

E12 >= E14

F12 >= F14

G12 >= G14

The *objective function* is to be set in cell C16 as = SUMPRODUCT(C3:G5, C9:G11).

In Solver the following settings are to be considered (see Figure 7.31). In the *Set Objective* we have the formula inserted in $C$16, which required to be minimized as we are looking to minimize the total transportation cost. Under the section *By Changing Variables Cells* select $C$9:$G$11. This is set for the 15 variables for this particular example. Within the section *Subject to the Constraints* the two sets of constraints are inserted. They are as follows:

Constraints set 1: $H$9:$H$11 <= $J$9:$J$11

Constraints set 2: $C$12:$G$12 >= $C$14:$G$14

Figure 7.31 Direct transportation example – Solver

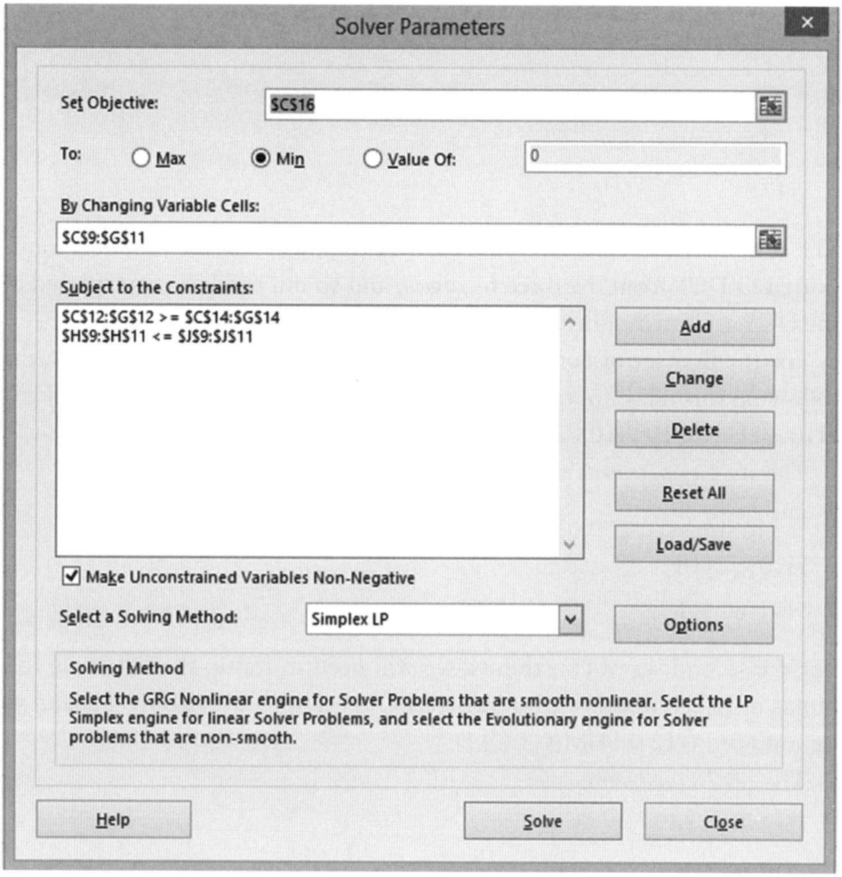

Figure 7.32 Direct transportation example – solution

	A	B	C	D	E	F	G	H	I	J
2		**Cost/unit**	**DC1**	**DC2**	**DC3**	**DC4**	**DC5**			
3		FactoryNT1	13	21	25	20	14			
4		FactoryNT2	10	24	11	26	27			
5		FactoryNT3	17	18	21	12	13			
6										
7	Distribution plan									
8		**Transportation Plan**	**DC1**	**DC2**	**DC3**	**DC4**	**DC5**	*Total volume transported*		*Plant Capacity*
9		FactoryNT1	250	0	0	0	0	250	<=	450
10		FactoryNT2	100	0	400	0	0	500	<=	500
11		FactoryNT3	0	245	0	200	150	595	<=	600
12		*Total received*	350	245	400	200	150			
13			>=	>=	>=	>=	>=			
14		*Customer Demand*	350	245	400	200	150			
15										
16		*Total Transportation cost*	17410							

The results are indicated in Figure 7.32. The total transportation cost identified in this case is 17410. To meet demand at the distribution centres, products are being transported only from FactoryNT1 to DC1. In this case, 250 units can be transported to DC1 from the 450 available. FactoryNT1 will experience products left over of 200 units.

From FactoryNT2, products are transported to DC1 and DC3. The entire capacity has been utilized in this case. From FactoryNT3 products are being transported to three locations – DC2, DC4 and DC5 – where the total number of products transported is 595, just a little below the number of products available in stock.

Demand at each distribution centre has been satisfied in full.

The model indicates that an overall capacity of 205 units is still left at the source, where this could be transferred into demand, or captured within production planning with the evaluation that this volume is not to be included.

Observations. A further discussion that is to take place in this instance is the mode of transport to be used to deliver the products.

For instance, let us assume that DC4 and DC5 are sufficiently close to FactoryNT3. If the distance to travel from FactoryNT3 to DC3 and back to the FactoryNT3 plus the distance to transport products from the FactoryNT3 to DC5 and back is higher in value than the distance to travel from FactoryNT3 to DC4 and DC5, then the transport manager at the FactoryNT3 may decide to carry out the transportation of the products in one go. However, this can only be achieved if there is sufficient capacity in the transport vehicle considered for this transportation.

Examples of different vehicle routing problems (VRP) will be related later in this chapter, but the direct transportation is one example that can be linked to another model using linear programming or a vehicle routing problem.

The direct transportation model has been presented by many authors, using various representation and combinations of this example. Some of these examples can be found in Albright and Winston (2011), Powell and Baker (2011), Russell and Taylor (2014), Taylor (2016) and others.

Transshipment model

The transshipment model is presented by many authors as a network flow model due to its network representation (Ragsdale, 2018) or it can be seen as a more complex version of the direct transportation model (Baker, 2016; Russell and Taylor, 2014). When working with a transshipment model, products are delivered to the final destination via an intermediate point(s). There are other variations of a transshipment model where some of the products are transported through a transshipment point, and some others are transported directly to the final destination, or they can be transported from an intermediate point to another intermediate point before they are sent to the final destination.

Transshipment Example 1

The following transshipment example is designed to work with two intermediate points, starting with three sources and six customers or final destination points.

The model can be represented as in Figure 7.33.

Figure 7.33 The network flow diagram for the transshipment model

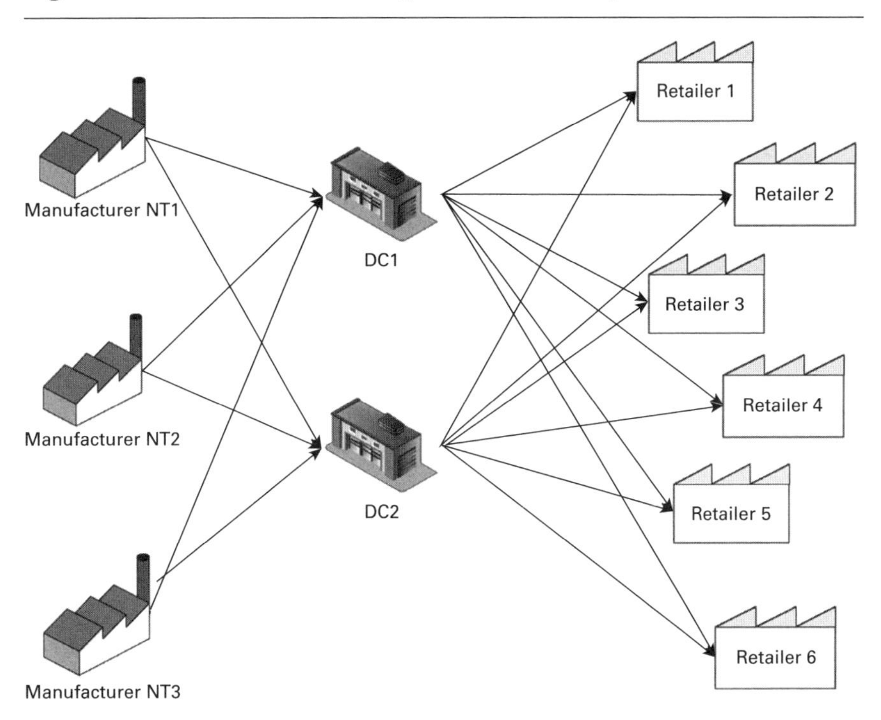

There are three manufacturing plants, named Manufacturer NT1, Manufacturer NT2 and Manufacturer NT3. Products can go through two distribution centres, named DC1 and DC2. They are then routed to the six retailers, named Retailer 1 (R1) to Retailer 6 (R6).

Input data. For this particular example there is no demand at the distribution centres and there are no capacity limits at these two distribution centres. Demand occurs at the retailers, and capacity is allocated to each manufacturing plant. Therefore, input data or data that must be collected by an analyst before determining the optimum plan for delivery, is as follows:

- *The number of organizations that are taking part in this delivery.* In this particular example we have three starting points or sources, two transshipment points and six destination points.
- *The capacity allocated to each manufacturing location*, or the sourcing.
- *The demand associated with each retailer/customer.*
- *The cost to transport products from one location to another.* In this case the cost of transporting products/unit from each of the three manufacturing plants to the two distribution centres and the cost of transporting products/unit from these two distribution centres to the six retailers.

Scope. The scope in this case is:

- Meet the customer demand by delivering products from the manufacturing plants to the two distribution centres and from the two distribution centres to the final destination by not exceeding the manufacturing plants capacity.
- Identify the minimum transportation cost when delivering these products through the network.
- Provide a delivery plan.

Output data. The expected output data at the end of the analysis is as follows:

- *Quantity to be delivered from each manufacturing plant to each distribution centre and from each distribution centre to each retailer.*
- *Total quantity delivered from each manufacturing plant and the leftover capacity at the manufacturing plant after a delivery.*
- *Total transportation cost.*
- *Total number of products delivered to each retailer.*
- *Total number of products that go through each distribution centre.*

The *mathematical model* in this case is as follows.

Variables. The number of products delivered from the three manufacturing plants to the two DCs and the number of products delivered from the two DCs to the six retailers. In total there are 18 variables.

Constraints. For this type of problem there are three sets of constraints: capacity constraints at the source; demand constraints from customers; and transshipment point constraints. To present these for our example we have the following:

Constrains set 1: The total number of products delivered from the manufacturing plants (sources) to the distribution centres (transshipment points) is less than or equal to (<=) the manufacturing plants' capacities.

As in our example there are three manufacturing plants, we will expect to see three defined constraints such as:

- *Total number of products delivered from Manufacturer NT1 to DC1+ total number of products delivered from Manufacturer NT1 to DC2 <= Manufacturer NT1 capacity (6,000 units).*
- *Total number of products delivered from Manufacturer NT2 to DC1+ total number of products delivered from Manufacturer NT2 to DC2 <= Manufacturer NT2 capacity (5,500 units).*
- *Total number of products delivered from Manufacturer NT3 to DC1+ total number of products delivered from Manufacturer NT3 to DC2 <= Manufacturer NT3 capacity (6,000 units).*

Constrains set 2: The total number of products delivered from the two distribution centres to the six retailers is higher than or equal to (>=) the customer demand.

As there are six retailers we are expecting to see six individual inequalities presented in this case such as:

- *Total number of products delivered from DC1 to R1 + total number of products delivered from DC2 to R1 >= retailer 1 demand (1,500).*
- *Total number of products delivered from DC1 to R2 + total number of products delivered from DC2 to R2 >= retailer 2 demand (1,250).*
- *Total number of products delivered from DC1 to R3 + total number of products delivered from DC2 to R3 >= retailer 3 demand (1,550).*

- *Total number of products delivered from DC1 to R4 + total number of products delivered from DC2 to R4 >= retailer 4 demand (2,100).*
- *Total number of products delivered from DC1 to R5 + total number of products delivered from DC2 to R5 >= retailer 5 demand (3,500).*
- *Total number of products delivered from DC1 to R6 + total number of products delivered from DC2 to R6 >= retailer 6 demand (5,000).*

Constrains set 3: The total number of products delivered from the manufacturing plants to the two distribution centres is equal to (=) the total number of products delivered from the distribution centres to the six retailers. As there are only two distribution centres in our case, two equations are expected to this set of constraints such as:

- *Total number of products from Manufacturer NT1 to DC1+ total number of products from Manufacturer NT2 to DC1 + total number of products from Manufacturer NT3 to DC1 = total number of products delivered from DC1 to R1 + total number of products delivered from DC1 to R2 + total number of products delivered from DC1 to R3 + total number of products delivered from DC1 to R4 + total number of products delivered from DC1 to R5 + total number of products delivered from DC1 to R6.*
- *Total number of products from Manufacturer NT1 to DC2 + total number of products from Manufacturer NT2 to DC2 + total number of products from Manufacturer NT3 to DC2 = total number of products delivered from DC2 to R1 + total number of products delivered from DC2 to R2 + total number of products delivered from DC2 to R3 + total number of products delivered from DC2 to R4 + total number of products delivered from DC2 to R5 + total number of products delivered from DC2 to R6.*

Objective function. The objective function in this case is to *minimize the total transportation cost*, which is formed of the total transport cost to deliver products from the three manufacturing plants to the two distribution centres plus the total cost to deliver products from the two distribution centres to the six retailers.

A template can be constructed in Excel (Figure 7.34).

Figure 7.34 Transshipment example 1 – Excel template

	A	B	C	D	E	F	G	H	I
1		cost	DC1	DC2					
2		M NT1	2	3					
3		M NT2	3	4					
4		M NT3	2	3					
5									
6		cost	R1	R2	R3	R4	R5	R6	
7		DC1	1.2	1.5	4	3	2	1.7	
8		DC2	1.7	1.9	2.3	2.1	3.2	3	
9									
10									
11			DC1	DC2	Products sent to DCs		Capacity		
12		M NT1	0	0	0	<=	6000		
13		M NT2	0	0	0	<=	5500		
14		M NT3	0	0	0	<=	4750		
15		Products delivered to the DCs	0	0					
16									
17									
18			R1	R2	R3	R4	R5	R6	Products delivered from the DCs
19		DC1	0	0	0	0	0	0	0
20		DC2	0	0	0	0	0	0	0
21		Products delivered	0	0	0	0	0	0	
22			>=	>=	>=	>=	>=	>=	
23		Demand	1500	1250	1550	2100	3500	5000	
24									
25									
26	Total cost	0							

The input values are observed as the cost of transporting one unit from the three manufacturers to the two distribution centres in the first table, whereas in the second table there is the cost of transporting products from the two distribution centres to the six retailers. The second part of the template represents the setting of the *variables* within the two tables from C12 to D14 for the total number of products delivered from the three manufacturing plants to DC1 and DC2. Also, from C19 to H20 the second set of variables are set, which will represent the total number of products sent from the DC1 and DC2 to the six retailers.

Setting *constrains set 1* in Excel requires the use of the following formulas:

In E12 = SUM(C12:D12); E13 = SUM(C13:D13) and E14 = SUM(C14:D14). We will then have:

E12 <= G12

E13 <= G13

E14 <= G14

Setting *constrains set 2* in Excel requires to use of the following formulas: C21 = SUM(C19:C20); D21 = SUM(D19:D20); E21 = SUM(E19:E20): F21 = SUM(F19:F20); G21 = SUM(G19:G20); H21 = SUM(H19:H20). Then we will have:

C21 >= C23

D21 >= D23

E21 >= E23

F21 >= F23

G21 >= G23

H21 >= H23

Setting *constrains set 3* in Excel requires to use of the following formulas: in cell C15 = SUM(C12:C14) and in cell D15 = SUM(D12:D14).

In cell I19 = SUM(C19:H19) and in cell I20 = SUM(C20:H20)

Then we will have

C15 = I19 and

D15 = I20

The *objective function* is to be set in B26 = SUMPRODUCT(C2:D4,C12:D14) + SUMPRODUCT(C7:H8, C19:H20)

All these formulas are to be inserted in Solver as indicated in Figure 7.35.

In *Set Objective* we have the formula inserted in $B$26, that required to be minimized as we are looking to minimize the total transportation cost. Under the section *By Changing Variables Cells* select $C$12:$D$14,$C$19:$H$20. In total we can note that there are 18 cells selected. Within the section *Subject to the Constraints* the three sets of constraints are inserted. They are as follows:

Figure 7.35 Transshipment example 1 – Solver

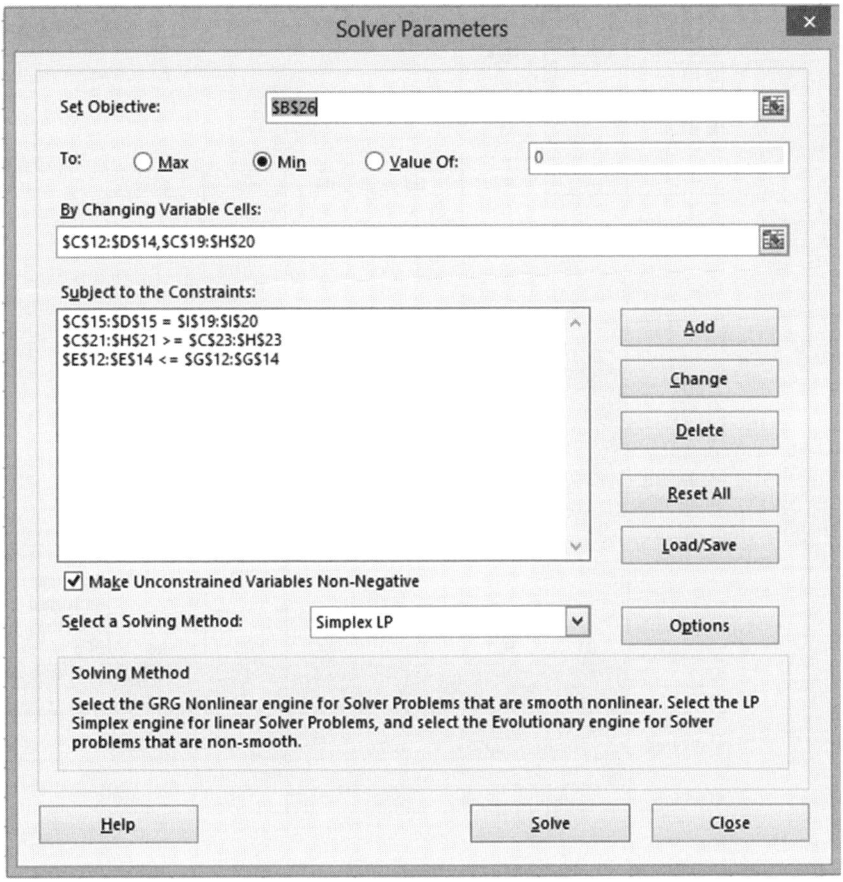

Solver Parameters

Set Objective: $B$26

To: Max / Min (selected) / Value Of: 0

By Changing Variable Cells:
$C$12:$D$14,$C$19:$H$20

Subject to the Constraints:
$C$15:$D$15 = $I$19:$I$20
$C$21:$H$21 >= $C$23:$H$23
$E$12:$E$14 <= $G$12:$G$14

Add | Change | Delete | Reset All | Load/Save

[x] Make Unconstrained Variables Non-Negative

Select a Solving Method: Simplex LP | Options

Solving Method

Select the GRG Nonlinear engine for Solver Problems that are smooth nonlinear. Select the LP Simplex engine for linear Solver Problems, and select the Evolutionary engine for Solver problems that are non-smooth.

Help | Solve | Close

Constraints set 1: $E$12:$E$14 <= $G$12:$G$14

Constraints set 2: $C$21:$H$21 >= $C$23:$H$23

Constraints set 3: $C$15:$D$15 = $I$19:$I$20

Following from the set expected output data, the results in Figure 7.36 have been obtained.

Figure 7.36 Transshipment example 1 – solution

	A	B	C	D	E	F	G	H	I
1		cost	DC1	DC2					
2		M NT1	2	3					
3		M NT2	3	4					
4		M NT3	2	3					
5									
6		cost	R1	R2	R3	R4	R5	R6	
7		DC1	1.2	1.5	4	3	2	1.7	
8		DC2	1.7	1.9	2.3	2.1	3.2	3	
9									
10									
11			DC1	DC2	Products sent to DCs		Capacity		
12		M NT1	4450	1550	6000	<=	6000		
13		M NT2	4150	0	4150	<=	5500		
14		M NT3	4750	0	4750	<=	4750		
15		Products delivered to the DCs	13350	1550					
16									
17									
18			R1	R2	R3	R4	R5	R6	Products delivered from the DCs
19		DC1	1500	1250	0	2100	3500	5000	13350
20		DC2	0	0	1550	0	0	0	1550
21		Products delivered	1500	1250	1550	2100	3500	5000	
22			>=	>=	>=	>=	>=	>=	
23		Demand	1500	1250	1550	2100	3500	5000	
24									
25									
26	Total cost	64540							

The quantities to be delivered from each manufacturing plant to each distribution centre are displayed in E12, E13 and E14. We note that the entire number of products allocated to M NT1 (6,000 units) has been delivered to DC1 (4,450 units) and DC2 (1,550 units), therefore the entire capacity has been used. However, we can also note an imbalance between the transportation loads to the two DCs. This resulted from the higher cost to transport products to DC2. From M NT2 only 4,150 products have been delivered, and a total of 1,350 are still left in stock at M NT2. The entire number of products has been transported from M NT3 to DC1.

The quantity of products delivered from each distribution centre to each retailer can be noted in cells C21 to H21, where it is evident that the quantities deliver do meet the customer demand in full. Numerically, this has been expected, as the total capacity available for delivery is higher than the total demand. If demand were higher than the total capacity, this would result in not meeting the customer demand in full. We can also note in this case that products from DC1 are transported to R1, R2, R4, R5 and R6, whereas DC2 only transports products to R3. At this point we can note again an imbalance of activities between the two distribution centres.

One other measure that we are looking for is the total number of products that go through each distribution centre. We have 13,350 products delivered through DC1 and 1,550 products delivered through DC2. In this particular example we have not set any capacities to the DC(s); however, if there are capacity limitations at these points, these figures would change.

The total transportation cost for the entire activity resulted in a value of £64,540. We can also note the split cost between each stage of transportation, such as it costs £3,550 to transport products from the manufacturing plants to the two DC(s) and it costs £29,040 to transport products from the DC(s) to the six retailers.

Observations. This type of example can be used in various applications in the distribution of goods for local as well as global operations. This also has applications for various modes of transport that not only consider road transportation but could also look at other modes, such as air, rail and maritime transport planning. Multi-model transportation can also be combined in this model.

Similar examples of transshipment can be found in Powell and Baker (2011), Taylor (2016) and Baker (2016).

A number of variations can be considered to this example, such as: the number of facilities forming the network; the links between each facility can change; the number of products being distributed can also change; capacity at each facility can be considered, and others. Another example that works with a variation of the links between facilities is presented below.

Transshipment example 2

This example aims to extend the transshipment model with two additional conditions. For this example, products can be transported from the sources to the intermediate points, the transshipment points, and they can be transported directly from the sources to customers. The second condition added in this case is the use of capacity to the transshipment points.

Therefore, let us assume the example, to start with, considers two factories (NTF1 and NTF2), four distribution centres (DC1, DC2, DC3 and DC4) and five customers (C1, C2, C3, C4 and C5).

The network is represented in Figure 7.37, where only two sets of links have been displayed to allow for clarity of representation. It should be noted, however, that links between all nodes in the system can be established.

Figure 7.37 Transshipment example 2

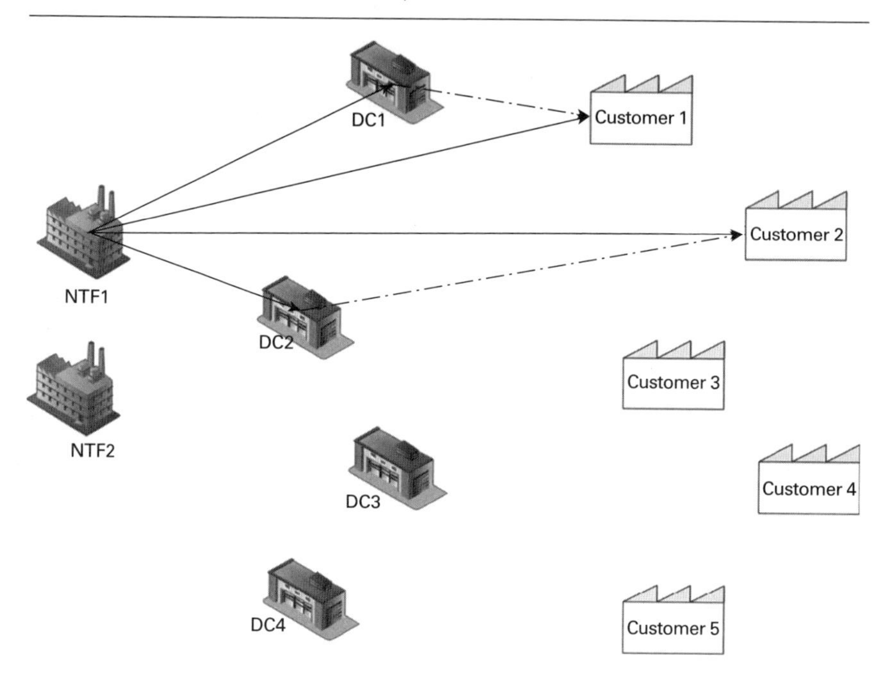

Input data. The input data, or data that must be collected by an analyst before determining the optimum plan for delivery, is as follows:

- *Number of organizations taking part in this delivery.* In this particular example we have two starting points or sources, four transshipment points and five destination points. In total this model will operate with 11 organizations in a two-stage transportation model.
- *Capacity allocated to each factory location.*
- *Capacity allocated to each distribution centre.*
- *Demand associated with each customer considered in this particular model*

- *Cost to transport products from one location to another.* In this case the cost of transporting products/unit from each of the two factories to the four distribution centres, the cost of transporting products/unit from the four distribution centres to the five customers, and the cost of transporting products/unit from factories to the five customers.

Scope. The scope of the model in this case is:

- Meet the customer demand by delivering products from the factories to the four distribution centres and from the four distribution centres to the final destination, final customers, while not exceeding the factory capacity.
- Identify the minimum transportation cost when delivering these products through the distribution network.
- Provide a delivery plan.
- Collect performance measurement information that would allow for an improvement in the decision-making process.

Output data. The expected output data at the end of the analysis is as follows:

- *Quantity to be delivered from each factory to each distribution centre and from each distribution centre to each customer and from each factory to each customer.*
- *Total quantity of leftover capacity at the factories after delivery.*
- *Total quantity of leftover capacity at the distribution centres after delivery.*
- *Total transportation cost at each stage in the process.*
- *Overall total transportation cost.*
- *Total number of products delivered to each customer.*
- *Total number of products that go through each distribution centre.*

The mathematical model in this case is as follows.

Variables. The number of products delivered from the two factories to the four DCs (8 variables) plus the number of products delivered from the four DCs to the five customers (20 variables) plus the number of products transported from the two factories to the five customers (10 variables). In total there are *38 variables*.

The template to represent this example is shown in Figure 7.38.

Figure 7.38 Transshipment example 2 – template

	A	B	C	D	E	F	G
1							
2	Cost/unit						
3		DC1	DC2	DC3	DC4		
4	NTF1	£1.50	£0.50	£1.20	£1.00		
5	NTF2	£1.50	£2.00	£1.80	£2.00		
6							
7		C 1	C 2	C 3	C 4	C 5	
8	NTF1	£3.00	£2.50	£3.00	£2.00	£1.50	
9	NTF2	£5.00	£4.00	£3.00	£3.00	£2.00	
10							
11		C 1	C 2	C 3	C 4	C 5	
12	DC1	£1.50	£1.50	£2.00	£1.50	£3.00	
13	DC2	£1.00	£1.50	£1.75	£1.00	£0.50	
14	DC3	£3.00	£1.50	£2.00	£2.00	£2.00	
15	DC4	£2.50	£4.00	£2.00	£1.75	£0.50	
16							
17							
18	Delivery Plan						
19							
20		DC1	DC2	DC3	DC4	Total from NTF(s) to DC(s)	
21	NTF1					0	
22	NTF2					0	
23	Total to DC(s) from NTF(s)	0	0	0	0		
24		<=	<=	<=	<=		
25	DCs capacity	105000	800000	120000	95000		
26							
27		C 1	C 2	C 3	C 4	C 5	Total from NTF(s) to C(s)
28	NTF1						0
29	NTF2						0
30	Total to C(s) from NTF(s)	0	0	0	0	0	
31							
32		Total from NTF(s)		NTF(s) Capacity			
33	NTF1	0	<=	750000			
34	NTF2	0	<=	800000			
35							
36		C 1	C 2	C 3	C 4	C 5	Total
37	DC1						0
38	DC2						0
39	DC3						0
40	DC4						0
41	Total from DC(s) to C(s)	0	0	0	0	0	
42							
43	Total to customer(s)	0	0	0	0	0	
44		>=	>=	>=	>=	>=	
45	Customer demand	90000	123000	105000	332000	356000	
46							
47	Total cost	£0.00					

Constraints. Three set of constraints were required for the previous example. In this example we have capacity allocated to the distribution centres, therefore one other set of constraints will be required.

Constrains set 1: The total number of products delivered from the factories (NTF1 and NTF2) to the four distribution centres (DC1, DC2, DC3 and DC4) plus the total number of products delivered from the factories (NTF1 *and NTF2) to the five customers (C1, C2, C3, C4 and C5) is less than or equal to (<=) the factory capacities.*

As in this example we have:

- *Total number of products delivered from NTF1 to DC1+ total number of products delivered from NTF1 to DC2 + total number of products delivered from NTF1 to DC3 + total number of products delivered from NTF1 to DC4 + total number of products delivered from NTF1 to C1+ total number of products delivered from NTF1 to C2 + total number of products delivered from NTF1 to C3 + total number of products delivered from NTF1 to C4 + total number of products delivered from NTF1 to C5 <= NTF1 capacity (750,000 units)*
- *Total number of products delivered from NTF2 to DC1+ total number of products delivered from NTF2 to DC2 + total number of products delivered from NTF2 to DC3 + total number of products delivered from NTF2 to DC4 + total number of products delivered from NTF2 to C1+ total number of products delivered from NTF2 to C2 + total number of products delivered from NTF2 to C3 + total number of products delivered from NTF2 to C4 + total number of products delivered from NTF2 to C5 <= NTF2 capacity (800,000 units)*

Constrains set 2: The total number of products delivered from the four distribution centres to the five customers plus (+) the total number of products delivered from the two factories to the five customers is higher than or equal to (>=) the customer demand. As there are five customers we expect to see five individual inequalities presented in this case, such as:

- *Total number of products delivered from DC1 to C1 + total number of products delivered from DC2 to C1 + total number of products delivered from DC3 to C1 + total number of products delivered from DC4 to C1+ total number of products delivered from NTF1 to C1 + total number of products delivered from NTF2 to C1 >= C1 demand (90,000 units).*
- *Total number of products delivered from DC1 to C2 + total number of products delivered from DC2 to C2 + total number of products delivered from DC3 to C2 + total number of products delivered from DC4 to C2 + total number of products delivered from NTF1 to C2 + total number of products delivered from NTF2 to C1 >= C2 demand (123,000 units).*
- *Total number of products delivered from DC1 to C3 + total number of products delivered from DC2 to C3 + total number of products delivered*

from DC3 to C3 + total number of products delivered from DC4 to C3+ total number of products delivered from NTF1 to C3 + total number of products delivered from NTF2 to C3 >= C3 demand (105,000 units).

- *Total number of products delivered from DC1 to C4 + total number of products delivered from DC2 to C4 + total number of products delivered from DC3 to C4 + total number of products delivered from DC4 to C4 + total number of products delivered from NTF1 to C4 + total number of products delivered from NTF2 to C4 >= C4 demand (332,000 units).*
- *Total number of products delivered from DC1 to C5 + total number of products delivered from DC2 to C5 + total number of products delivered from DC3 to C5 + total number of products delivered from DC4 to C5+ total number of products delivered from NTF1 to C5 + total number of products delivered from NTF2 to C5 >= C5 demand (356,000 units).*

Constrains set 3: The total number of products delivered from the two factories to the four distribution centres is equal to (=) the total number of products delivered from the distribution centres to the five customers. As there are four distribution centres in our case, four equations are expected for this set of constraints. To shorten the representation of these equation for the total number of products delivered, the notation 'total no of prod' will be used:

- *Total no of prod from NTF1 to DC1 + total no of prod from NTF2 to DC1 = total no of prod from DC1 to C1 + total no of prod from DC1 to C2 + total no of prod from DC1 to C3 + total no of prod from DC1 to C4 + total no of prod from DC1 to C5.*
- *Total no of prod from NTF1 to DC2 + total no of prod from NTF2 to DC2 = total no of prod from DC2 to C1 + total no of prod from DC2 to C2 + total no of prod from DC2 to C3 + total no of prod from DC2 to C4 + total no of prod from DC2 to C5.*
- *Total no of prod from NTF1 to DC3 + total no of prod from NTF2 to DC3 = total no of prod from DC3 to C1 + total no of prod from DC3 to C2 + total no of prod from DC3 to C3 + total no of prod from DC3 to C4 + total no of prod from DC3 to C5.*
- *Total no of prod from NTF1 to DC4 + total no of prod from NTF2 to DC4 = total no of prod from DC4 to C1 + total no of prod from DC4 to C2 + total no of prod from DC4 to C3 + total no of prod from DC4 to C4 + total no of prod from DC4 to C5.*

Constrains set 4: Total number of products delivered to the distribution centres is <= the capacity of the DC(s). As there are four DC(s) in this example, we expect to have four equations in this case:

- *Total no of prod from NTF1 to DC1 + total no of prod from NTF2 to DC1 <= DC1 capacity (105,000 units).*
- *Total no of prod from NTF1 to DC2 + total no of prod from NTF2 to DC2 <= DC2 capacity (800,000 units).*
- *Total no of prod from NTF1 to DC3 + total no of prod from NTF2 to DC3 <= DC3 capacity (120,000 units).*
- *Total no of prod from NTF1 to DC4 + total no of prod from NTF2 to DC4 <= DC4 capacity (95,000 units).*

Objective function. The objective function in this case is to *minimize the total transportation cost*, which is formed of the total transport cost to deliver products from the factories (NTF1 and NTF2) to the four DC(s) + the total cost to deliver products from the factories (NTF1 and NTF2) to the five customers + the total cost to deliver products from the four distribution centres to the five customers.

Solution. The model's details presented above have to be inserted in the template developed in Figure 7.38 and in Solver to be in a position to work with the optimum answer obtained in this case.

All input data needs to be clearly represented in the developed template. The cost of transporting products/unit is provided in the first part of the template from cells B4 to F15. Within the second part of the template the variables are being set as from cell B21 to E22; these are the number of products transported from NTF1 and NTF2 to the distribution centres. These are counted as eight variables. From cell B28 to cell F29 the number of products delivered from factories directly to the five customers are set. These result in 10 variables. The remaining 20 variables are those set from cell B37 to F40 and they are the number of products delivered from the four distribution centres to the five customers.

To set up the constraints we need to insert the required formulas in the template developed.

To indicate the sum of the total number of products delivered from the NFT1 to the four distribution centres use the formula = SUM(B21:E21) in cell F21, and the same goes for the products delivered from NTF2, in cell F22 = SUM(B22:E22).

In cells B23 to E23 we will have the sum of products delivered from the factories to each individual distribution centre such as:

In cell B23 = SUM(B21:B22); and in C23 = SUM(C21:C22); and in D23 = SUM(D21:D22) and in E23 = SUM(E21:E22). These formulas will be used to indicate the total number of products going into the transshipment point from factories.

In cell G28 and G29 we are going to calculate the sum of the total number of products delivered from the factories directly to customers. So, in G28 = SUM(B28:F28) and G29 = SUM(B29:F29).

From cell B30 to F30, we are also calculating the sum of the total number of products from the factories to individual customer as such: in cell B30 = SUM(B28:B29) and following the same logic up to cell F30 = SUM(F28:F29).

Within the following section we are interested to collect the information regarding the total number of products sent from each individual factory, to help us later on to set the constraint that the total number of products delivered from the factories is less than or equal to the factory capacity. Therefore, in cell B33 = F21 + G28 and in cell B34 = F22 + G29.

Within cell G37 to cell G40 the total sum of products from the DC(s) to individual customers is being collected such as in cell G37 = SUM(B37:F37) up to G40 = SUM(B40:F40). From cell B41 to F41 the sum of products sent from all DC(s) to each individual customer is being calculated as such B41 = SUM(B37:B40); C41 = SUM(C37:C40) and so on up to F41 = SUM(F37:F40).

From cells B43 to F43 we will be collecting the information regarding the total number of products delivered to each individual customer, as this will help us to set up the second main constraint referring to the total number of products delivered to customers (or destination) is higher than or equal to the customer demand. Therefore, in cell B43 = B30 + B41 and so on up to cell F43 = F30 + F41

The formula used for the objective function that indicates the minimum transportation cost for all possible routes is set in cell B47 = SUMPRODUCT (B4:E5,B21:E22) + SUMPRODUCT(B8:F9,B28:F29) + SUMPRODUCT(B12:F15,B37:F40)

Having developed this template and inserted all these formulas, we are now ready to set up the Solver. This will be as in Figure 7.39.

Figure 7.39 Transshipment example 2 – Solver

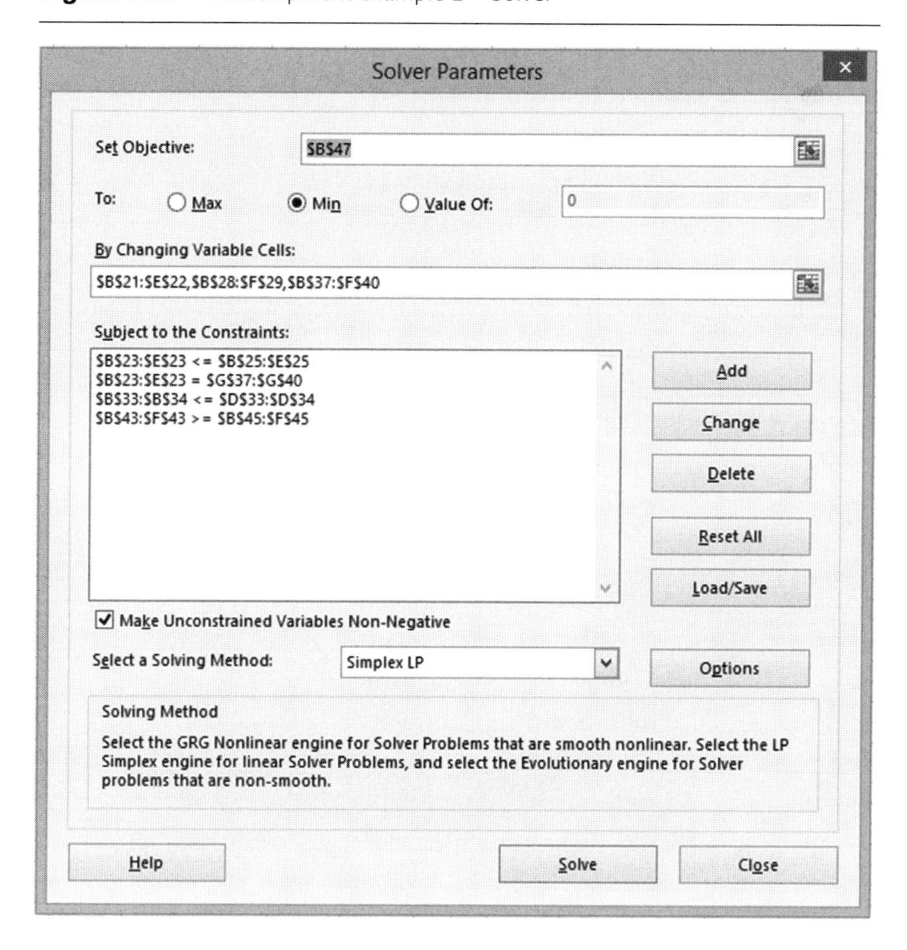

In *Set Objective* we have the formula inserted in $B$47, which required to be minimized as we are looking to minimize the total transportation cost. Under the section *By Changing Variable Cells* select the three sets of variables such as $B$21:$E$22,$B$28:$F$29,$B$37:$F$40. In total 38 variables are to be selected for this particular example. Within the section *Subject to the Constraints* the four sets of constraints are inserted.

Constraints set 1: $B$33:$B$34 <= $D$33:$D$34

Constraints set 2: $B$43:$F$43 >= $B$45:$F$45

Constraints set 3: $B$23:$E$23 = $G$37:$G$40

Constraints set 4: $B$23:$E$23 <= $B$25:$E$25

By solving this example the solution shown in Figure 7.40 can be noted.

Figure 7.40 Transshipment example 2 – solution

	A	B	C	D	E	F	G
16							
17							
18	Delivery Plan						
19							
20		DC1	DC2	DC3	DC4	Total from NTF(s) to DC(s)	
21	NTF1	0	750000	0	0	750000	
22	NTF2	105000	0	0	0	105000	
23	Total to DC(s) from NTF(s)	105000	750000	0	0		
24		<=	<=	<=	<=		
25	DCs capacity	105000	800000	120000	95000		
26							
27		C 1	C 2	C 3	C 4	C 5	Total from NTF(s) to C(s)
28	NTF1	0	0	0	0	0	0
29	NTF2	0	0	105000	0	46000	151,000
30	Total to C(s) from NTF(s)	0	0	105000	0	46000	
31							
32		Total from NTF(s)		NTF(s) Capacity			
33	NTF1	750,000	<=	750000			
34	NTF2	256,000	<=	800000			
35							
36		C 1	C 2	C 3	C 4	C 5	Total
37	DC1	0	105000	0	0	0	105000
38	DC2	90000	18000	0	332000	310000	750000
39	DC3	0	0	0	0	0	0
40	DC4	0	0	0	0	0	0
41	Total from DC(s) to C(s)	90,000	123,000	0	332,000	310,000	
42							
43	Total to customer(s)	90,000	123,000	105,000	332,000	356,000	
44		>=	>=	>=	>=	>=	
45	Customer demand	90000	123000	105000	332000	356000	
46							
47	Total cost	£1,701,000.00					
48							

Note that products are only sent through DC1 and DC2, where DC3 and DC4 are not being used. The product flow diagram is presented in Figure 7.41. Note also that factory NTF1 does not deliver directly to customers, where NTF2 delivers directly to C3 and C5, but also uses DC1. DC1 only delivers products to C2 where DC2 delivers products to C1, C2, C4 and C5.

Figure 7.41 Transshipment example 2 – product flow solution

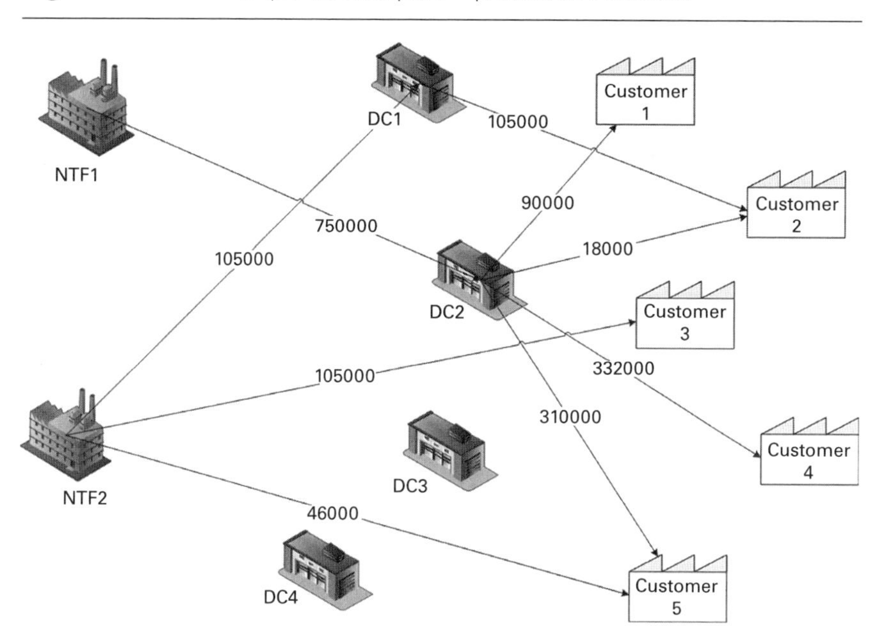

The NTF2 capacity is not used in full, and this indicates that maybe more demand from customers could make better use of the products produced at this factory. Customer demand, however, is met in full, which again indicates that attracting more customers may be one strategic point for this supply chain distribution.

The total transportation cost in this case is £1,701,000. A lot can be learned from this modelling and interesting strategic decisions could be considered in this case.

Observations. This type of model can also indicate which facilities are not being used, such as DC3 and DC4 in the example above.

This model can also give indications on locating facilities closer to their operations. For example, in the case studied, it can be noted that C2 is being served by both distribution centres (DC1 and DC2). Therefore, if possible, the location of any of these facilities may be reduced to reduce the distance travelled.

Facility location problem

The facility location problem plays a key role in the supply chain, and the network design of the supply chain. Establishing the location of a facility is a very important decision, as this facility is not easy to move; therefore this may need to be seen as a long-term decision. There are a number of factors that impact the network design decision in relation to facility allocation in the supply chain.

Some of these are *strategic factors* that may be looking at creating a competitive advantage in deciding the position of a facility such as the location of a warehouse or a manufacturing plant. This may also be creating a low-cost strategy, where the facility location would be determined by bringing it to a low-cost location, or a location where raw materials are much lower in cost, or the manufacturing cost is low. It could also be a decision based on convenience, so a facility may be located very close to customers, for example convenience stores such as Sainsbury's Local or Tesco Express, which are smaller stores located very close to customers. One other strategic aspect is product availability, where large stores are located at a remote location where customers need to travel a longer distance, however due to the size of the store the range of products available to purchase is larger.

One other factor in deciding the location of a facility is *technology and infrastructure*. For example, a decision may be made to locate a warehouse closer to a motorway to facilitate faster transportation and avoid city traffic, and to have access to other sources such as electricity and water supply. There is also the issue of not having access to good infrastructure, which will result in higher transportation costs as well as longer lead times for transporting to and from the facility located in such place. There are also *economic factors*, where organizations may prefer to locate a factory in a country where they may negotiate to open job opportunities to local employees. Exchange rate could be another factor that may influence the network design decision.

Clustering also plays an important role in facility location. Organizations may prefer to locate a manufacturing plant or a retail facility within a cluster, where resources can be shared, access and infrastructure is already in place and demand has already been generated by the other members in the cluster. Access to a larger market share is another incentive for facility location.

The scope of the model in this case is to maximize the profitability of the supply chain network while at the same time being in a position to keep a high customer service and meet customer demand. When designing the best location of a facility, a number of trade-offs need to be taken into consideration between cost and responsiveness. An organization may decide to build a number of warehouses in different locations to reach as many customers as possible, which should generate a high response rate. However, the cost of building and maintaining a larger number of warehouses will be high.

The decision elements that need to be taken into consideration for these models are: the location and capacity of the facility; meeting demand at the facility; and the distance to travel to and from this facility.

To present a model that deals with the best location for a facility, the following example will be used. Let us assume that we are interested in finding the best location for a distribution centre when delivering to 10 main customers in the region. We know the customers' location, given as x and y coordinates. From past experience, we also know the annual demand of shipments.

Figure 7.42 gives details about the x and y coordinates for each customer and annual shipment.

Figure 7.42 DC location – input data

Customers	X-coordinate	Y-coordinate	Annual shipment
C1	18	–5	234
C2	7	13	350
C3	–8	21	100
C4	4	–21	500
C5	32	23	600
C6	14	11	750
C7	24	–10	255
C8	–3	4	148
C9	1	28	589
C10	16	24	488

Input data:

- *Customer address* – and with this we will know their position on the map. In this example we have the x and y coordinate for the set of customers.
- *Annual shipment.* This detail needs to be collected before starting the analysis. This information is in general available from previous historical data, or it can be predicted.

Goal. Identify the best location for the distribution centre (DC) location from which products are being transported to each customer.

Objective function. The total distance travelled from the new DC location to each customer per year to be minimized.

Output data:

- *Location of the DC as x and y coordinates.*
- *Distance travelled from customers to the DC.*
- *Total annual distance travelled.*

The template shown in Figure 7.43 can be created in Excel for this type of problem. Each customer can be displayed on a graph (or located on a map) following the details from the x and y coordinates.

To calculate the distance between the new DC location and each customer, Pythagoras' formula will be used, for example:

$$d(C1,\ DC) = \sqrt{(x_{C1} - x_{DC})^2 + (y_{C1} - y_{DC})^2}$$

In Excel the formula :=SUMXMY2(x Range, y Range) can be used for calculating $(x_{c1} - x_{dc})^2 + (y_{c1} - y_{dc})^2$

Therefore, for the distance between C1 and DC we will have in cell E6 we can insert: =SQRT(SUMXMY2(B6:C6,$B$17:$C$17)). The same formula can be followed for all customers' coordinates, for example for the distance between DC and C10 we will have in cell E15: =SQRT(SUMXMY2(B15:C15,$B$17:$C$17))

To calculate the total annual distance from each customer to the new DC location, the formula to be used in cell E17 is: =SUMPRODUCT(D6:D15,E6:E15). This formula is being used as we assume that the distance travelled from DC location to customers is being multiplied with the annual shipment. In reality, this formula will be adjusted depending on the mode of transport used and the capacity of the vehicle considered for transportation. Still, to demonstrate this example, the simplified formula to be used is a sum of the annual shipment per customers multiplied with the total distance travelled.

We do not have variables in this case, but Solver can be used to identify the coordinates for the new DC location. In this case the x and y coordinated for the new DC location will be set in the *Changing Cells* section in Solver (see Figure 7.44).

After running Solver, the solution obtained in this case is provided in Figure 7.45.

Figure 7.43 DC location – Excel template

	A	B	C	D	E
1	**Distribution Centre Location**				
2					
3	*Input Data*				
4					
5	**Customers**	**X-coordinate**	**Y-coordinate**	**Annual Shipment**	**Distance from DC to Customers**
6	C1	18	-5	234	
7	C2	7	13	350	
8	C3	-8	21	100	
9	C4	4	-21	500	
10	C5	32	23	600	
11	C6	14	11	750	
12	C7	24	-10	255	
13	C8	-3	4	148	
14	C9	1	28	589	
15	C10	16	24	488	
16					
17	DC location			Total annual distance	
18					
19					
20					
21					
22					

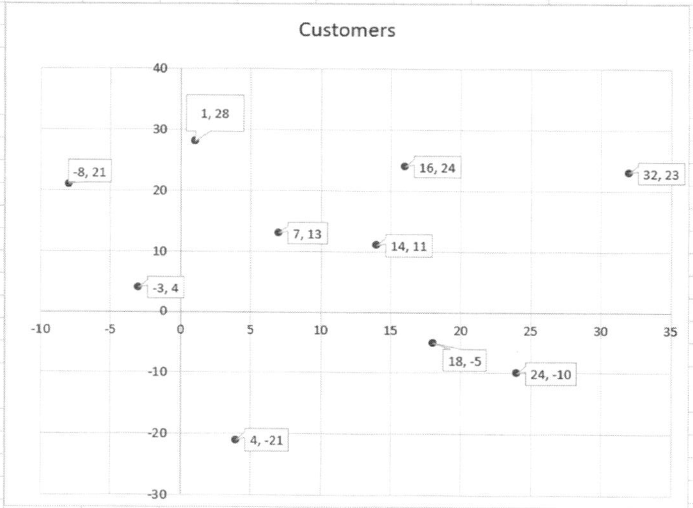

Figure 7.44 DC location – Solver

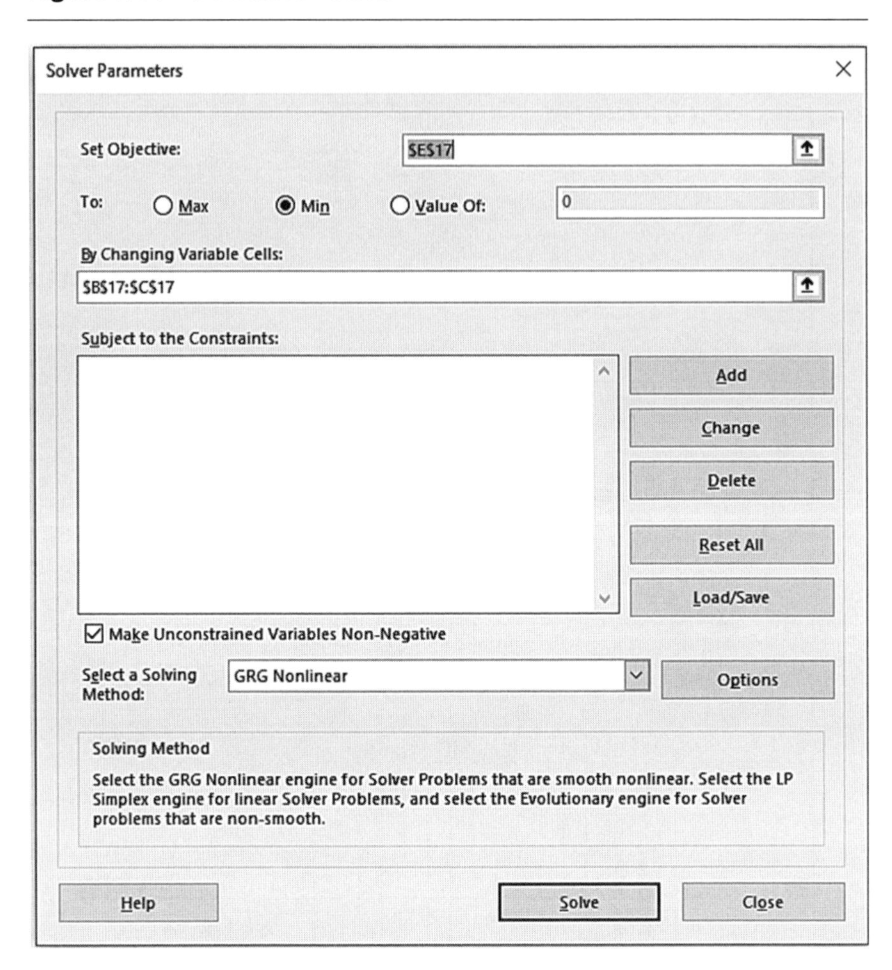

It can be noted that the proposed location for the new DC that will serve all customers on the list as well as minimize the total distance travelled will have the coordinates (14.00, 11.00). It can be noted in this case that the solution indicates the DC location should be built at the same coordinates as C6, the customer that also has the largest annual shipment per year.

Observations: A number of changes can be considered for this type of problem, one being the number of customers considered for analysis. The same evaluation can be considered for a larger number of customers as well as for just a few customers. The formulas used and the modelling process will remain the same regardless of the number of customers considered for the analysis.

Figure 7.45 DC location – solution

	A	B	C	D	E	F	G	H	I	J	K	L	M	N	O
1	**Distribution Centre Location**														
2															
3															
4															
5	**Customers**	**X-coordinate**	**Y-coordinate**	**Annual Shipment**	**Distance from DC to Customers**										
6	C1	18	-5	234	16.49										
7	C2	7	13	350	7.28										
8	C3	-8	21	100	24.17										
9	C4	4	-21	500	33.53										
10	C5	32	23	600	21.63										
11	C6	14	11	750	0.00										
12	C7	24	-10	255	23.26										
13	C8	-3	4	148	18.38										
14	C9	1	28	589	21.40										
15	C10	16	24	488	13.15										
16															
17	DC location	14.00	11.00	Total annual distance	66242.80										
18															
19															
20															
21															

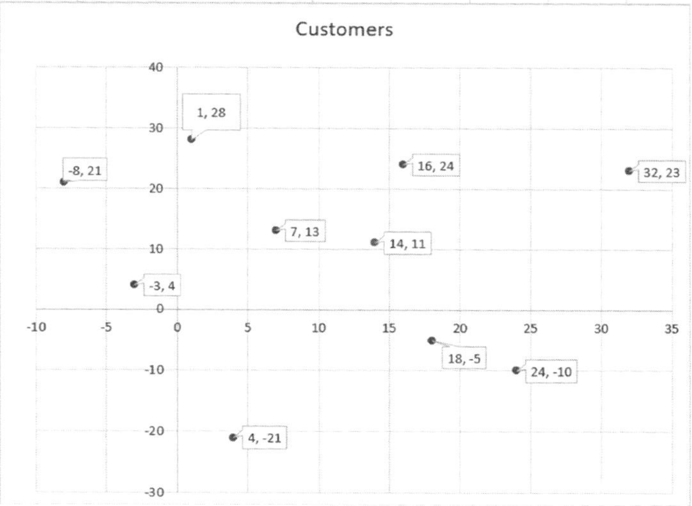

In addition to a change in the number of customers, one other change that can be considered is a change in the values considered for the annual shipment for each individual customer. A change in these values may result in a change in the answer obtained for the DC location, therefore different scenarios of potential different values should be considered before taking the final decision on where this DC should be located.

A change in the mode of transport will generate a change in the calculation of the total annual distance travelled. If air transportation is used, it can be assumed that the transportation will follow a straight line as opposed to using road transport where the calculation of the distance travelled will not be possible in a straight line, as roads are not built in straight lines. However, in many cases the distance between locations is already known, therefore calculation of the distance information will not be required.

There is also the situation when the answer provided may not be located on land, or on a location that is easily accessible, therefore decisions on location in this case will be made closer to the obtained answer.

Similar examples of this type of model are also discussed in Albright and Winston (2005).

Vehicle routing problem

Models solving the vehicle routing problem (VRP) are also captured and discussed in this chapter. These are not optimization models, as they will not necessarily provide the optimum solution. They are *heuristics* models that may have more than one solution that managers may consider when implementing these techniques in practice.

Vehicle routing algorithms have been published since 1959 with a paper from Dantzig and Ramser (1959) followed very shortly by another publication by Clarke and Wright (1964), and subsequently there has been a large number of publications dealing with this problem. In 2009, Laporte summarized a number of the VRP algorithms in 'Fifty years of vehicle routing'. The interest in this type of problem not only comes from its practical application, but also from the complexity of optimizing such problem. Two of these algorithms (savings algorithm and sweep algorithm) are discussed below.

'Savings algorithm' in the vehicle routing problem

In order to formulate the savings algorithm, as introduced by Clarke and Wright (1964), the following assumptions are considered. From a starting point that can be named as 'depot' products (or raw materials or semi-finished products) are delivered to customers based on their demand. To transport these products or goods, it can be assumed that a fixed number of capacitated vehicles are available at the depot.

For the example discussed in this particular case (see Figure 7.46), capacity is considered for each vehicle used. Vehicles are also restricted by the total distance they can travel on any one route. The routes covered by a vehicle will start and end at the starting point, at the depot.

The aim of the problem described here is to ensure that each customer is visited at least once, their demand is satisfied in full and the total distance travelled to visit all expected customers in a particular time period is minimized.

The vehicle's capacity cannot be exceeded, but there may be routes where vehicles will not travel to full capacity.

This is a heuristic model that will have more than one solution, and therefore transport managers can select any based on their requirements. The intention is to find a set of solutions that meet all the requirements.

This model can be solved using dedicated vehicle routing software packages. However, when the number of customers to deliver to is manageable, the problem can be solved using spreadsheets, where in this case the proposed solution is to use Excel.

To exemplify the savings, we can consider the example presented in Figure 7.46.

Figure 7.46 Vehicle routing – option 1

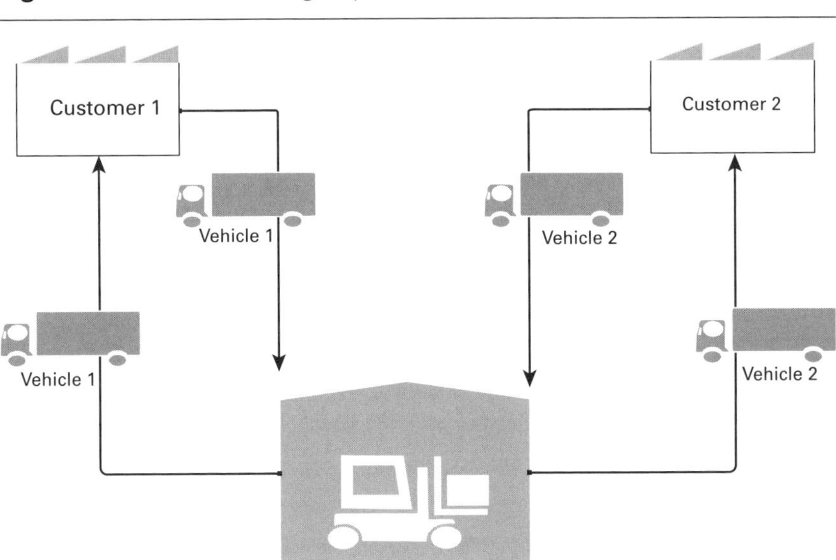

For two customers (C1 and C2) we can assume delivery of products using two vehicles (V1 and V2). If we consider the notation for the distance to be travelled from Depot (D) to C1 as d(D,C1) that is the same with the distance to be travelled from C1 to depot (D), it can be assumed that V1 travels d(D,C1) + D(C1, D) = 2*d(D, C1)

Following the same logic, we will have the distance travelled by vehicle 2 (V2) as 2*d(D, C2).

Therefore, the total distance travelled in this scenario is:

Total distance 1 = 2*d(D, C1) + 2*d(D, C2) (see Figure 7.46)

However, if there is sufficient capacity in any one of the vehicles, let us assume we select Vehicle 1, to carry the demand from C2, the new total distance travelled is:

> Total distance 2 = d(D, C1) + d(C1, C2) + d(C2, D). As presented in Figure 7.47, vehicle V1 starts the route at depot D, travels to customer C1, followed by customer C2 and returns back to depot D.

To note if any savings exist by using the round trip (see option 2, Figure 7.47), the different between Total distance 1 – Total distance 2 should be a positive value.

Figure 7.47 Vehicle routing – option 2

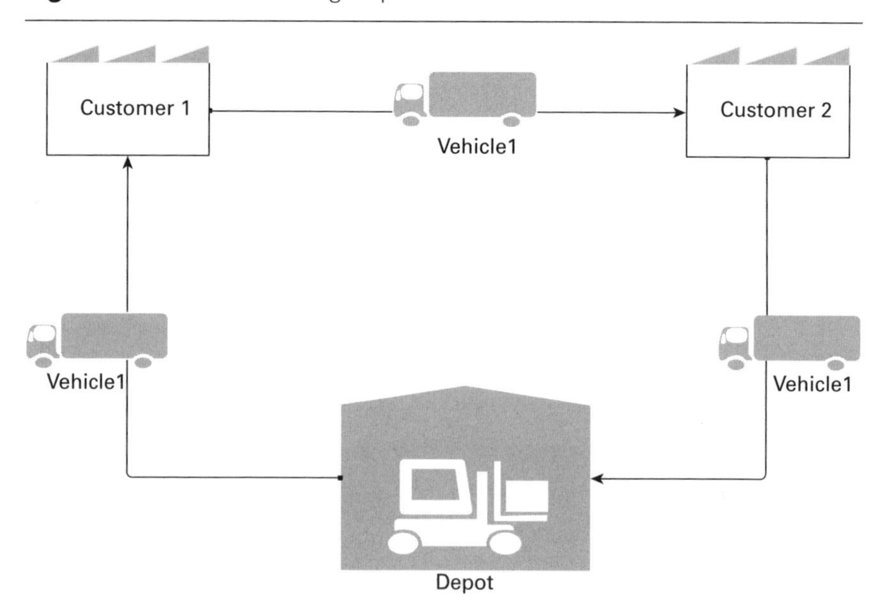

Therefore to define the savings between customer C1 and customer C2 we will have the following formula for savings s(C1,C2):

s(C1, C2) = 2*d(D, C1) + 2*d(D, C2) – d(D, C1) – d(C1, C2) – d(C2, D) therefore,

s(C1, C2) = d(D, C1) + d(D, C2) – d(C1, C2)

This savings formula will be used later on in the calculations.

Scope. The scope of the model is:

- Develop a routing plan that minimizes the total distance travelled by delivering to all customers on time and in full and by not exceeding the vehicle capacity and the total maximum number of miles allowed to travel by each vehicle.

Constraints. The constraints of this problem are:

- The number of vehicles is limited.
- The total distance travelled is predefined.
- The total weight to be carried by a vehicle (in our case to total number of cases) is limited. This also refers to the vehicle capacity being limited.
- The total number of customers to deliver to is selected.

This type of problem is typical for home deliveries from a retailer or a distribution centre to retailers, or directly from a distribution centre or warehouse to customers. We also see this problem in last mile delivery examples. With more and more online delivery, this example is one that has a number of practical applications.

Input data. This is data that an analyst must obtain before the routing plan can be established.

- Total number of customers in any one particular time period who are placing an order. For this example let us assume that we have 15 customers (marked as C1, C2, ..., C15), who require deliveries in one routing plan. This routing plan can be considered as developed per day.
- The address of the customers and the address of the depot the products/goods are being delivered to/from to be collected and verified. In our case, we will consider an x and y coordinate for each customer, with the coordinates being (0, 0) for the depot. When collecting the delivery address, this is in general exemplified by the postcode and located on the most

up-to-date routing map. The distance between each point will then be calculated. However, in practice, this distance will be known based on the postcode of each customer.

- Customer demand, or the quantity required by each customer. In this example, cases will be used to represent the customers' demand.
- Total number of vehicles available.
- Total distance allowed for a vehicle for any one route.
- Total weight (number of cases) each vehicle can transport in any one route.

The input data for our example is presented in Figure 7.48 where we have five vehicles available that can travel a maximum distance of 160 miles, and carry no more than 130 cases in total. For this particular example, 15 customers are considered, where for each customer the location is known. In this example the location is given as the (x, y) coordinates, where in reality this is linked to the postcode of their address, for which the longitude and latitude is consider to position the customer on the map. We also have the demand in this case, given as the number of cases required by each customer.

Solution. This model can be developed and solved using spreadsheets. This approach can be considered as soon as the number of customers is manageable. However, if there are a larger number of customers, a dedicated vehicle routing software package (such as Paragon) should be considered. The following steps could be considered to identify a number of feasible solutions.

Step 1. Identify all the customers, their location and demand that are to be included in the analysis. In this example, the identified customers are presented in Figure 7.48. Based on the customer address (in our case their (x, y) coordinates), the task is to plot the customers on the map. Within dedicated computer software such as Paragon for VRP, this step is identified as *Locate Customers*.

Step 2. Calculate the *Distance Matrix* for any possible combination of paring two customers, as well as between the customers and the depot. For this particular example, it is considered that the forward route is the same as the return route. However, it can be argued that in practice this may not be the case due to road restrictions, variations in speed limitations and any other traffic conditions. To calculate the distance matrix in this particular case, Pythagoras' formula will be used for the distance between two points.

$$d(C1,\ C2) = \sqrt{(x_{C1} - x_{C2})^2 + (y_{C1} - y_{C2})^2}$$

Figure 7.48 VRP savings algorithm – input data

No of available vehicles	5
Maximum allocated distance travel per vehicle/route	160 miles
Maximum number of cases allocated per vehicle/route	130 cases

Location	X-coordinate	Y-coordinate	Demand (Cases)
Depot	0	0	N/A
C1	4	11	10
C2	45	-30	40
C3	15	36	26
C4	-30	36	28
C5	-20	-46	50
C6	8	-35	30
C7	-12	38	23
C8	-35	20	35
C9	19	-17	11
C10	15	-7	20
C11	18	25	10
C12	-30	-15	25
C13	-25	-50	40
C14	38	-28	45
C15	20	0	15

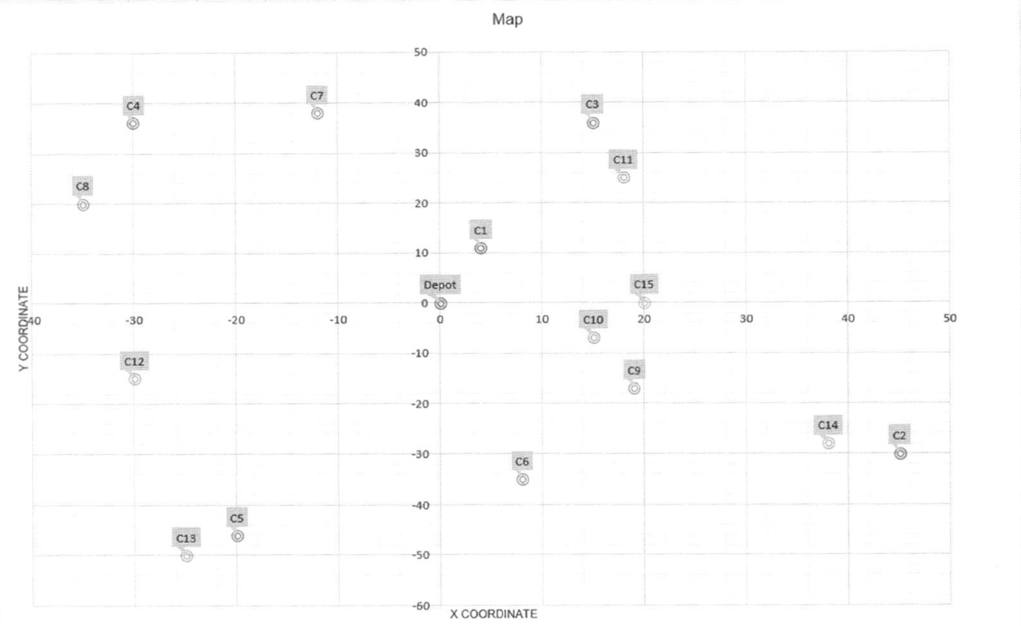

Where d(C1, C2) represents the distance between customer C1 and customer C2; *xC1, xC2* and *yC1, yC2* represent the coordinates for each location of customer C1 and C2 respectively.

A template is to be constructed in Excel for calculating the distance matrix. The formulas to be used in Excel are as follows.

To calculate the distance matrix in Excel, the following formula could be used for any combination for values inserted in the table. Therefore, for the distance between the depot and customer C1 we can use:

d(D, C1) = SQRT(SUMXMY2($C$7:$D$7,C6:D6)), where in cells C7:D7 are the coordinates for C1, and in cells C6:D6 are inserted the coordinates for the depot (see Figure 7.49). This formula can be copied for all the values in the same column, and this can be done by locking the cells C7 and D7 as $C$7:$D$7 followed by copy the formula up to cells C21: D21. Still, the formula will need to be inserted again for calculating the values for C2 and so on. So, for example for the

d(D,C2)= =SQRT(SUMXMY2($C$8:$D$8,C6:D6)), where the formula used to calculate the d(C15, C5) =SQRT(SUMXMY2($C$11:$D$11,C21:D21)) (see Figure 7.49).

Step 3. This now follows with the calculation of the *savings matrix* (see Figure 7.50). To calculate the savings, the formula demonstrated earlier will be used. For example, the savings between customer C1 and customer C2 follows the formula:

s(C1, C2) = d(D, C1) + d(D, C2) – d(C1, C2) where the savings between customer C1 and customer C6 will be calculated as:

s(C1, C6) = d(D, C1) + d(D, C6) – d(C1, C6)

Step 4. This step is concerned with the generation of the *Savings List*. The aim is to identify the groups of customers for which the highest savings can be achieved.

The *Savings Matrix* is the key instrument that allows for the construction of the *Savings List*. The *Savings List* is ordered with the highest savings value first and this will identify the group of customers with the highest savings.

For our example, we have the highest savings for the group of customers (C5, C13) with a saving of 99.66 miles, followed by the group (C2, C14) = 94.00 miles (see for more details Figure 7.51). This list finishes with a group of customers that indicates no savings, such as (C3, C5) = 0 savings, or with groups of customers such as (C1, C5) = 0.02, and (C8, C10) = 0.04 with very low savings. We will know from this list that there is no reason to group these customers in the same route, as they will not provide any savings. However, it is relevant to focus on the set of customers with the highest savings.

Figure 7.49 VRP savings calculations for the distance matrix

Maximum allocated distance travel per vehicle/route	160 miles
Maximum number of cases allocated per vehicle/route	130 cases

Location	X-coordinate	Y-coordinate	Demand (Cases)
Depot	0	0	N/A
C1	4	11	10
C2	45	-30	40
C3	15	36	26
C4	-30	36	28
C5	-20	-46	50
C6	8	-35	30
C7	-12	38	23
C8	-35	20	35
C9	19	-17	11
C10	15	-7	20
C11	18	25	10
C12	-30	-15	25
C13	-25	-50	40
C14	38	-28	45
C15	20	0	15

Distance Matrix	Depot	C1	C2	C3	C4	C5	C6	C7	C8	C9	C10	C11	C12	C13	C14	C15
Depot	=SQRT(SUMXMY2($C$7:$D$7,C6:D6))			39.00	46.86	50.16	35.90	39.85	40.31	25.50	16.55	30.81	33.54	55.90	47.20	20.00
C1	11.7[illegible]	SUMXMY2(array_x, array_y)		27.31	42.20	61.85	46.17	31.38	40.02	31.76	21.10	19.80	42.80	67.54	51.74	19.42
C2	54.08	57.98	0.00	72.50	99.90	66.94	37.34	88.73	94.34	29.07	37.80	61.27	76.49	72.80	7.28	39.05
C3	39.00	27.31	72.50	0.00	45.00	89.16	71.34	27.07	52.50	53.15	43.00	11.40	68.01	94.85	68.01	36.35
C4	46.86	42.20	99.90	45.00	0.00	82.61	80.53	18.11	16.76	72.18	62.24	49.24	51.00	86.15	93.38	61.61
C5	50.16	61.85	66.94	89.16	82.61	0.00	30.08	84.38	67.68	48.60	52.40	80.53	32.57	6.40	60.73	60.96
C6	35.90	46.17	37.34	71.34	80.53	30.08	0.00	75.69	69.81	21.10	28.86	60.83	42.94	36.25	30.81	37.00
C7	39.85	31.38	88.73	27.07	18.11	84.38	75.69	0.00	29.21	63.13	52.48	32.70	55.97	88.96	82.80	49.68
C8	40.31	40.02	94.34	52.50	16.76	67.68	69.81	29.21	0.00	65.46	56.82	53.24	35.36	70.71	87.37	58.52
C9	25.50	31.76	29.07	53.15	72.18	48.60	21.10	63.13	65.46	0.00	10.77	42.01	49.04	55.00	21.95	17.03
C10	16.55	21.10	37.80	43.00	62.24	52.40	28.86	52.48	56.82	10.77	0.00	32.14	45.71	58.73	31.14	8.60
C11	30.81	19.80	61.27	11.40	49.24	80.53	60.83	32.70	53.24	42.01	32.14	0.00	62.48	86.45	56.65	25.08
C12	33.54	42.80	76.49	68.01	51.00	32.57	42.94	55.97	35.36	49.04	45.71	62.48	0.00	35.36	69.23	52.20
C13	55.90	67.54	72.80	94.85	86.15	6.40	36.25	88.96	70.71	55.00	58.73	86.45	35.36	0.00	66.73	67.27
C14	47.20	51.74	7.28	68.01	93.38	60.73	30.81	82.80	87.37	21.95	31.14	56.65	69.23	66.73	0.00	33.29
C15	20.00	19.42	39.05	36.35	61.61	60.96	37.00	49.68	58.52	17.03	8.60	25.08	52.20	67.27	33.29	0.00

Figure 7.50 VRP savings calculations for the savings matrix

Distance Matrix	Depot	C1	C2	C3	C4	C5	C6	C7	C8	C9	C10	C11	C12	C13	C14	C15
Depot	0.00	11.70	54.08	39.00	46.86	50.16	35.90	39.85	40.31	25.50	16.55	30.81	33.54	55.90	47.20	20.00
C1	11.70	0.00	57.98	27.31	42.20	61.85	46.17	31.38	40.02	31.76	21.10	19.80	42.80	67.54	51.74	19.42
C2	54.08	57.98	0.00	72.50	99.90	66.94	37.34	88.73	94.34	29.07	37.80	61.27	76.49	72.80	7.28	39.05
C3	39.00	27.31	72.50	0.00	45.00	89.16	71.34	27.07	52.50	53.15	43.00	11.40	68.01	94.85	68.01	36.35
C4	46.86	42.20	99.90	45.00	0.00	82.61	80.53	18.11	16.76	72.18	62.24	49.24	51.00	86.15	93.38	61.61
C5	50.16	61.85	66.94	89.16	82.61	0.00	30.08	84.38	67.68	48.60	52.40	80.53	32.57	6.40	60.73	60.96
C6	35.90	46.17	37.34	71.34	80.53	30.08	0.00	75.69	69.81	21.10	28.86	60.83	42.94	36.25	30.81	37.00
C7	39.85	31.38	88.73	27.07	18.11	84.38	75.69	0.00	29.21	63.13	52.48	32.70	55.97	88.96	82.80	49.68
C8	40.31	40.02	94.34	52.50	16.76	67.68	69.81	29.21	0.00	65.46	56.82	53.24	35.36	70.71	87.37	58.52
C9	25.50	31.76	29.07	53.15	72.18	48.60	21.10	63.13	65.46	0.00	10.77	42.01	49.04	55.00	21.95	17.03
C10	16.55	21.10	37.80	43.00	62.24	52.40	28.86	52.48	56.82	10.77	0.00	32.14	45.71	58.73	31.14	8.60
C11	30.81	19.80	61.27	11.40	49.24	80.53	60.83	32.70	53.24	42.01	32.14	0.00	62.48	86.45	56.65	25.08
C12	33.54	42.80	76.49	68.01	51.00	32.57	42.94	55.97	35.36	49.04	45.71	62.48	0.00	35.36	69.23	52.20
C13	55.90	67.54	72.80	94.85	86.15	6.40	36.25	88.96	70.71	55.00	58.73	86.45	35.36	0.00	66.73	67.27
C14	47.20	51.74	7.28	68.01	93.38	60.73	30.81	82.80	87.37	21.95	31.14	56.65	69.23	66.73	0.00	33.29
C15	20.00	19.42	39.05	36.35	61.61	60.96	37.00	49.68	58.52	17.03	8.60	25.08	52.20	67.27	33.29	0.00

=$C$30+C31-D31

Savings Matrix	C1	C2	C3	C4	C5	C6	C7	C8	C9	C10	C11	C12	C13	C14	C15
C3	23.39	20.58	0												
C4	16.36	1.04	40.86	0											
C5	0.02	37.30	0.00	14.41	0										
C6	1.43	52.65	3.56	2.23	55.98	0									
C7	20.17	5.20	51.78	68.60	5.63	0.06	0								
C8	11.99	0.05	26.81	70.41	22.79	6.40	50.95	0							
C9	5.44	50.51	11.34	0.18	27.05	40.30	2.21	0.35	0						
C10	7.16	32.83	12.55	1.17	14.31	23.59	3.92	0.04	31.28	0					
C11	22.71	23.62	58.40	28.42	0.44	5.88	37.96	17.88	14.29	15.22	0				
C12	2.44	11.14	4.53	29.40	51.13	26.50	17.42	38.50	10.00	4.39	1.86	0			
C13	0.06	37.18	0.05	16.62	99.66	55.56	6.80	25.50	26.40	13.73	0.26	54.09	0		
C14	7.17	94.00	18.19	0.68	36.63	52.30	4.25	0.15	50.74	32.61	21.36	11.51	36.37	0	
C15	12.29	35.03	22.65	5.25	9.20	18.90	10.17	1.79	28.47	27.95	25.73	1.34	8.63	33.92	0

Figure 7.51 VRP savings calculations for the savings list

Savings list					
Customers group	**Savings**	**Customers groups**	**Savings**	**Customers groups**	**Savings**
C5, C13	99.66	C9, C13	26.40	C1, C10	7.16
C2, C14	94.00	C11, C15	25.73	C7, C13	6.80
C4, C8	70.41	C8, C13	25.50	C6, C8	6.40
C4, C7	68.60	C2, C11	23.62	C6, C11	5.88
C3, C11	58.40	C6, C10	23.59	C5, C7	5.63
C5, C6	55.98	C1, C3	23.39	C1, C9	5.44
C6, C13	55.56	C5, C8	22.79	C4, C15	5.25
C12, C13	54.09	C1, C11	22.71	C2, C7	5.20
C2, C6	52.65	C3, C15	22.65	C3, C12	4.53
C6, C14	52.30	C11, C14	21.36	C10, C12	4.39
C3, C7	51.78	C2, C3	20.58	C7, C14	4.25
C5, C12	51.13	C1, C7	20.17	C7, C10	3.92
C7, C8	50.95	C6, C15	18.90	C3, C6	3.56
C9, C14	50.74	C3, C14	18.19	C1, C12	2.44
C2, C9	50.51	C8, C11	17.88	C4, C6	2.23
C3, C4	40.86	C7, C12	17.42	C7, C9	2.21
C6, C9	40.30	C4, C13	16.62	C11, C12	1.86
C8, C12	38.50	C1, C4	16.36	C8, C15	1.79
C7, C11	37.96	C10, C11	15.22	C1, C6	1.43
C2, C5	37.30	C4, C5	14.41	C12, C15	1.34
C2, C13	37.18	C5, C10	14.31	C4, C10	1.17
C5, C14	36.63	C9, C11	14.29	C2, C4	1.04
C13, C14	36.37	C10, C13	13.73	C4, C14	0.68
C2, C15	35.03	C3, C10	12.55	C5, C11	0.44
C14, C15	33.92	C1, C15	12.29	C8, C9	0.35
C2, C10	32.83	C1, C8	11.99	C11, C13	0.26
C10, C14	32.61	C12, C14	11.51	C4, C9	0.18
C9, C10	31.28	C3, C9	11.34	C8, C14	0.15
C4, C12	29.40	C2, C12	11.14	C1, C13	0.06
C9, C15	28.47	C7, C15	10.17	C6, C7	0.06
C4, C11	28.42	C9, C12	10.00	C2, C8	0.05
C10, C15	27.95	C5, C15	9.20	C3, C13	0.05
C5, C9	27.05	C13, C15	8.63	C8, C10	0.04
C3, C8	26.81	C1, C2	7.81	C1, C5	0.02
C6, C12	26.50	C1, C14	7.17	C3, C5	0.00

Step 5. At this step the routes are constructed by grouping customers with the highest savings.

An initial consideration in this case will be to connect the customers with the highest savings, such as C5 with C13, C2 with C14, C4 with C8 and C4 with C7, C3 with C11. Just from this initial attempt, it can be observed that sections of the routes can already be observed on the graph.

If we start with the calculations for the route that incorporates the customers as follow: C8, C4, C7, C3 and C11, we can now question whether C1 should be allocated to this first route or not. So, this will depend not only on the restrictions of the distance travelled, but also on the vehicle capacity.

Therefore, route 1 can be defined as:

Route 1: D–C1–C11–C3–C7–C4–C8–D

For this route, the distance travelled will be: d(Route 1) = D29 + D40 + N32 + F36 + J33 + G37 + K29 = 145.15 miles and the total number of cases to be carried is: c(Route 1) = E7 + E17 + E9 + E13 + E10 + E14 = 127 cases.

Based on both of these conditions, the route obtained in this case is a viable route.

Continuing the same logic, we can see:

Route 2: D–C12–C13–C5–D

Distance travelled on route 2: d(Route 2) = O29 + H29 + O42 + P34 = 125.46 miles

Cases carried: c(Route 2) = E18 + E19 + E11 = 115 cases

This will appear in the developed template calculated in cell D48=$C$30+C31–D31 and is represented in Figure 7.50.

Route 3: D–C6–C2–C14–C9–D

Distance travelled: d(Route 3) = I29 + L29 + I31 + E43 + Q38 = 127.97 miles

Cases carried: c(Route 3) = E12 + E8 + E20 + E15 = 126 cases

At this point there are only two customers left to be allocated in a route: C10 and C15.

They will need to be grouped together in a final route; however, from the savings table they do not appear to be in the list of the groups with the highest savings. Still, considering that all customers are required to be visited a minimum of once, Route 4 could be developed as:

Route 4: D–C10–C15–D

Distance travelled: d(Route 4) = C39 + C44 + M44 = 45.16 miles

Cases carried: c(Route 4) = E16 + E21 = 35 cases.

Solution 1 is:

- Route 1: D–C1–C11–C3–C7–C4–C8–D, with d(Route1-S1) = 145.15 miles to travel, carrying c(Route1-S1) = 127 cases.
- Route 2: D–C12–C13–C5–D, with d(Route 2-S1) = 125.46 miles to travel, carrying c(Route 1-S1) = 115 cases.
- Route 3: D–C6–C2–C14–C9–D, with d(Route3-S1) = 127 miles to travel, carrying c(Route3-S1) = 126 cases.
- Route 4: D–C10–C15–D, with d(Route4-S1) = 45.16 miles to travel, carrying c(Route4-S1) = 35 cases.

Total distance travelled – Solution 1 – 443.75 miles.

Figure 7.52 VRP savings routes – solution 1

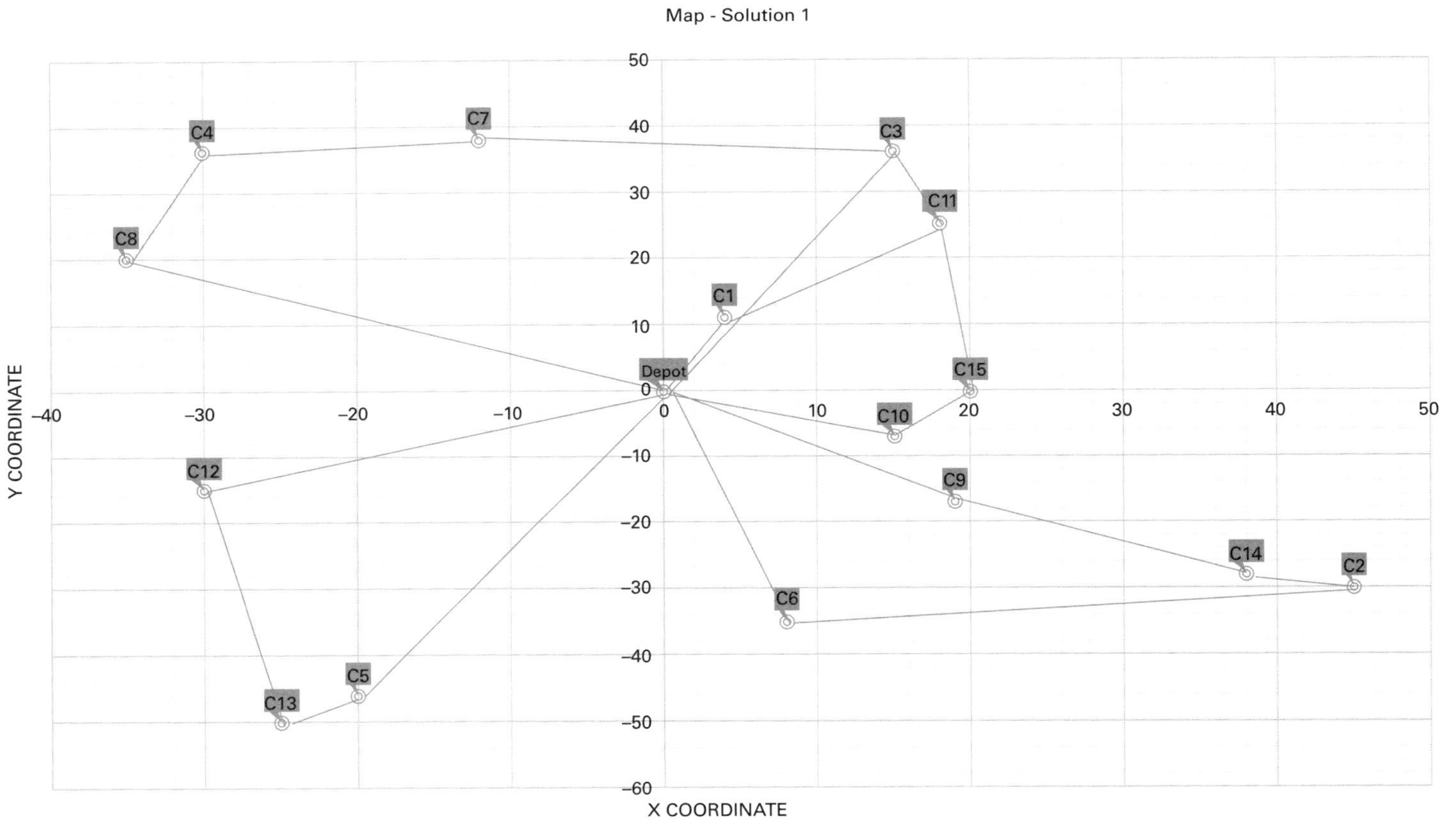

Therefore, we will have four routes, with the first three routes well balanced in terms of distance travelled and number of cases carried. Still, there is an imbalance generated by route 4 where the total miles travelled and the total number of cases carried is lower in comparisons with the other three routes.

This could be overcome by a different management decision, as to consider additional deliveries to other customers, or using a smaller vehicle; however, this limitation could be overcome by recalculating route 1.

Step 6. Alternative solutions. To identify an alternative solution the following can be considered.

The same steps are now repeated as for Solution 1; however, C1 will now be removed from Route 1 and reallocated to Route 4. In this case, there are no changes to Route 2 and Route 3, however for Route 1 the total distance travelled is now reduced to 133.45, to deliver to five customers, and the total number of cases is 117. Where for Route 4, the total distance travelled has now increased to 67.67, with the number of cases of 45 to be delivered to three customers.

Solution 2 (Figure 7.53):

- Route 1: D–C11–C3–C7–C4–C8–D, with d(Route1-S2) = 144.47 miles to travel, carrying c(Route1-S2) = 117 cases.
- Route 2: D–C12–C13–C5–D, with d(Route2-S2) = 125.46 miles to travel, carrying c(Route2-S2) = 115 cases.
- Route 3: D–C6–C2–C14–C9–D, with d(Route3-S2) = 127 miles to travel, carrying c(Route3-S2) = 126 cases.
- Route 4: D–C10–C15–C1–D, with d(Route4-S2) = 67.67 miles to travel, carrying c(Route4-S2) = 45 cases.

Total distance travelled – Solution 2 – 465.57 miles.

Solution 2 provides a more balanced routing plan, if this is required. However, the overall distance travelled in Solution 1 is 443.75 miles, which is lower than in Solution 2, 465.57 miles. In this situation, the calculations provide alternatives, although the final decision is a management decision that needs to meet the overall goals of the organization.

Other alternatives that could be incorporated in these manual calculations are the time require to load/unload and delivery time windows.

The search for a new solution can continue, and in this case, we would like to explore if C11 is transferred to Route 4.

Figure 7.53 VRP savings routes – solution 2

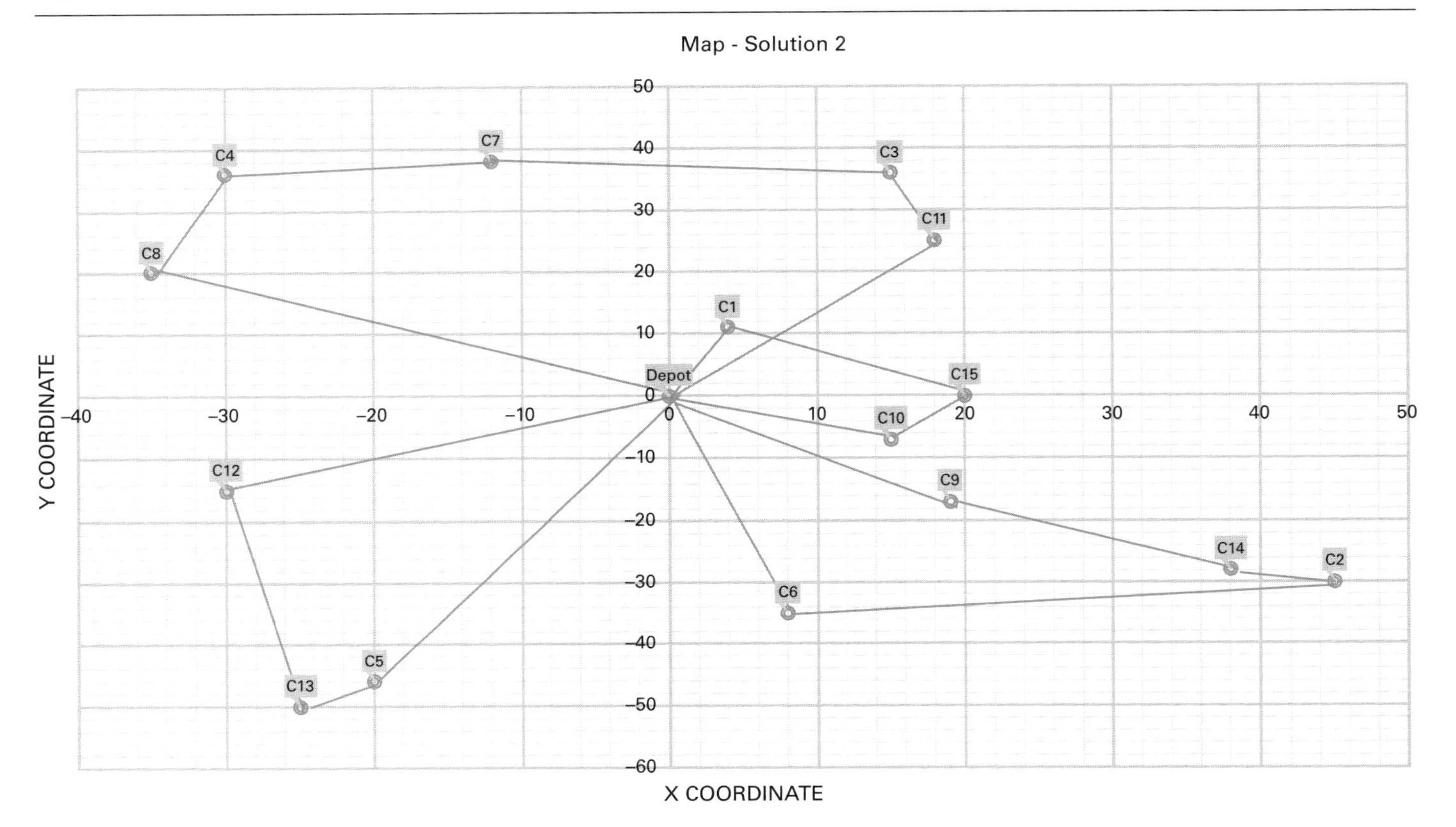

Therefore, we are going to move the investigation to determine Solution 3 for this routing plan as follows.

Solution 3:

- Route 1: D–C3–C7–C4–C8–D, with d(Route1-S3) = 141.26 miles to travel, carrying c(Route1-S3) = 107 cases.
- Route 2: D–C12–C13–C5–D, with d(Route2-S3) = 125.46 miles to travel, carrying c(Route2-S3) = 115 cases.
- Route 3: D–C6–C2–C14–C9–D, with d(Route3-S3) = 127 miles to travel, carrying c(Route3-S3) = 126 cases.
- Route 4: D–C10–C15–C11–C1–D, with d(Route4-S3) = 81.74 miles to travel, carrying c(Route4-S3) = 55 cases.

Total distance travelled – Solution 3 – 476.43 miles (Figure 7.54).

The balanced of delivering to 3 to 4 customers per route has been improved considering Solution 3, but the overall distance travelled is now higher than in Solutions 1 and 2. Therefore, depending on the routing goal, the best solution can be calculated to meet the transport organization's requirements.

The search for alternative solutions can continue, for example a split of Route 1 and Route 2 can be considered as the savings between C8, C12 is significant at 38.50 miles. Therefore, we could explore another solution below.

Solution 4:

- Route 1: D–C7–C4–C8–C12–D, with d(Route1-S4) = 143.62 miles to travel, carrying c(Route1-S4) = 106 cases.
- Route 2: D–C13–C5–C6–D, with d(Route2-S4) = 128.29 miles travelled, carrying c(Route2-S4) = 120 cases.
- Route 3: D–C9–C14–C2–C10–D, with d(Route3-S4) = 109.08 miles travelled, carrying c(Route3-S4) = 116 cases.
- Route 4: D–C15–C11–C3–C1–D, with d(Route4-S4) = 95.50 miles travelled, carrying c(Route4-S4) = 61 cases.

The overall distance travelled in this case is again 476.49 miles, as obtained for Solution 3.

Therefore, it can be concluded that Solution 4 (Figure 7.55) is viable, but the overall distance travelled in this case is higher than in Solution 1.

Figure 7.54 VRP savings routes – solution 3

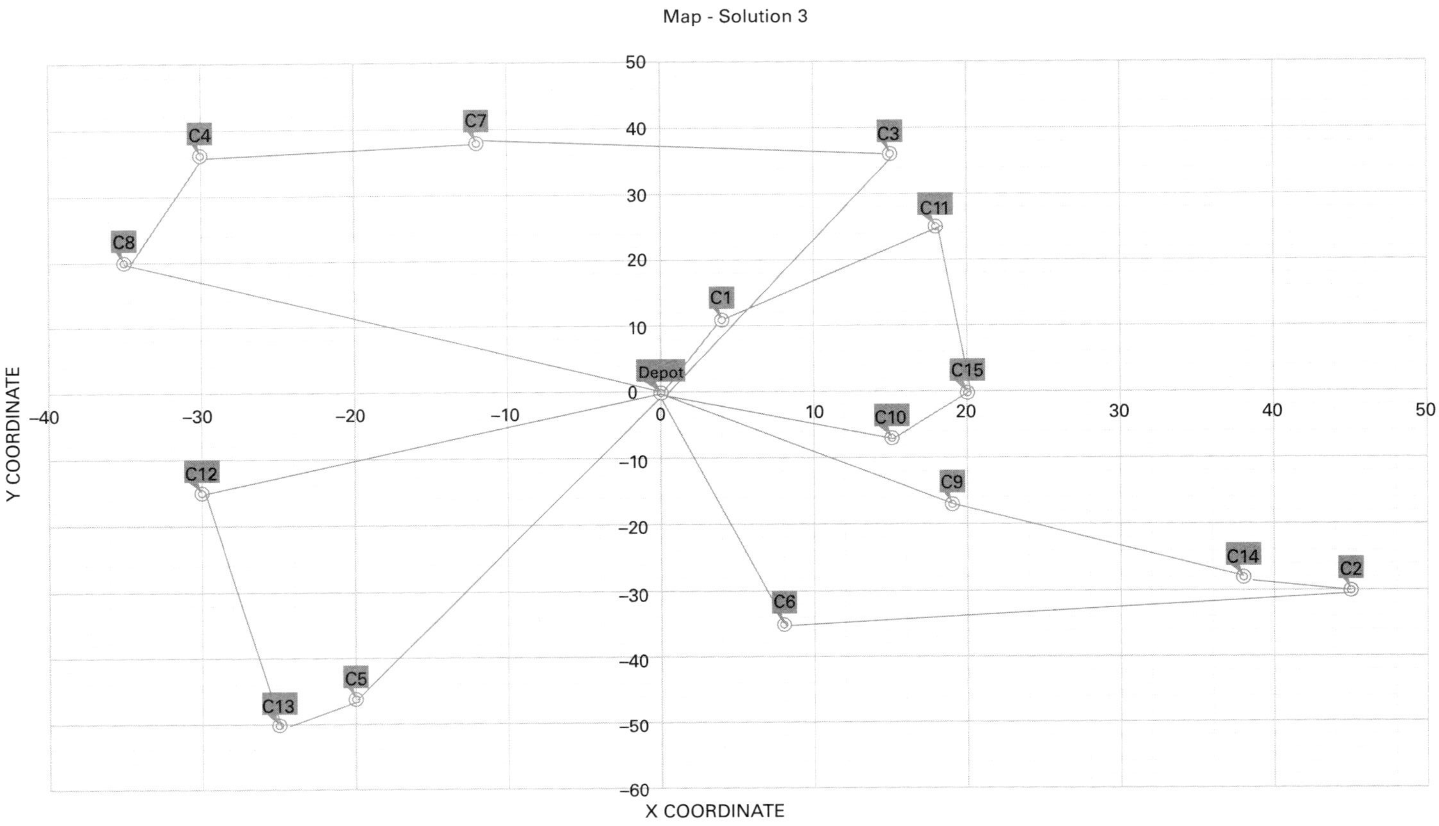

Map - Solution 3

Depot
C1
C2
C3
C4
C5
C6
C7
C8
C9
C10
C11
C12
C13
C14
C15

X COORDINATE

Y COORDINATE

Figure 7.55 VRP savings routes – solution 4

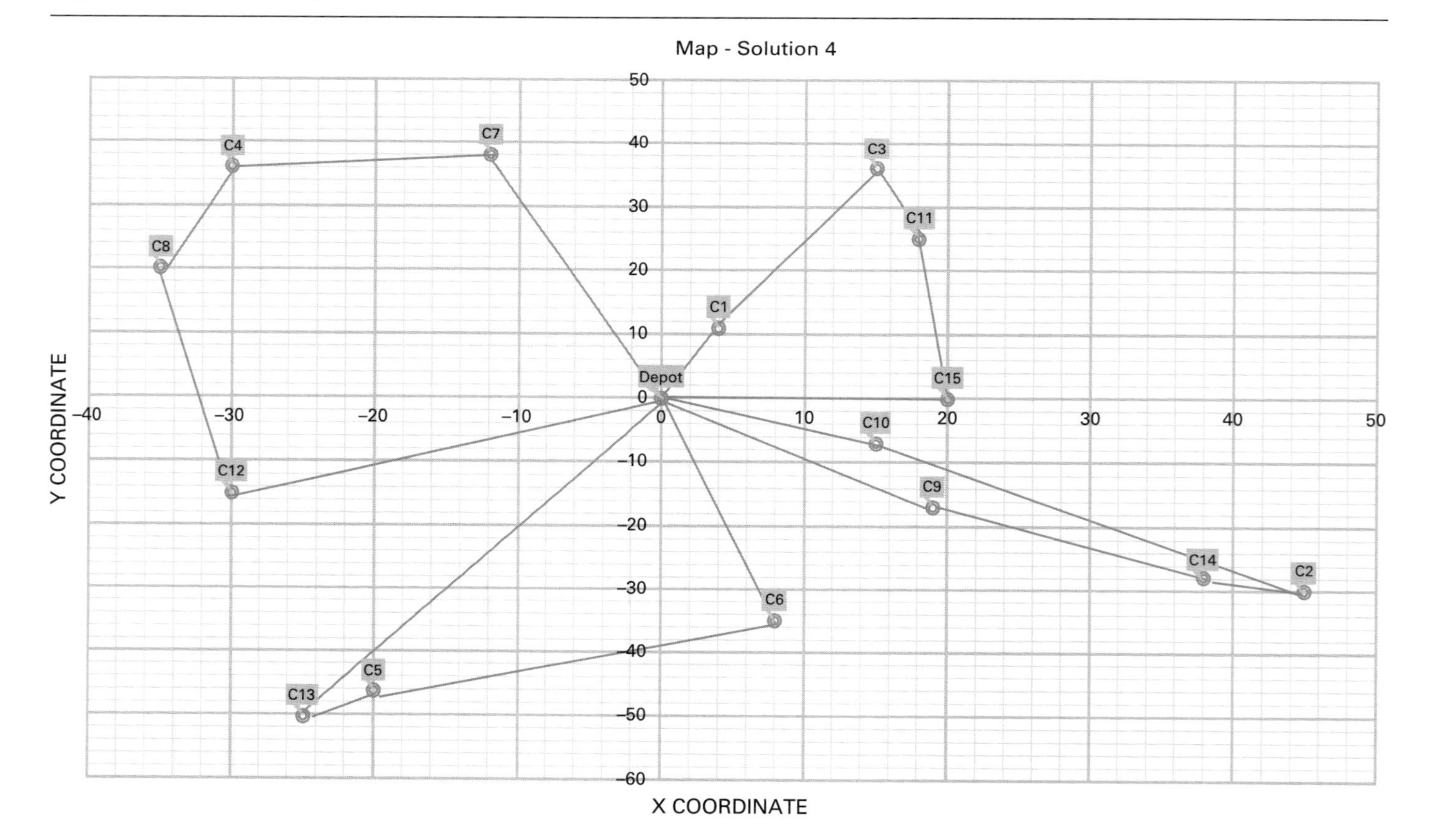

A number of other solutions can be explored in this case, following the *Savings List* to guide the most appropriate grouping for determining a route.

When looking to apply this technique in practice, a number of other details may also need to be added, such as a particular time window of delivery, road speed restrictions, congestion time, vehicle size and vehicle availability; and particular product characteristics, such as perishable products, will have delivery restrictions in relation to time and temperature. These conditions can be considered with the savings algorithm model before determining the routes; however, they can also be incorporated at the end of the calculated route plans.

Sweep algorithm in the vehicle routing problem

Looking at a modification of the previously presented algorithm, the sweep algorithm will consider some of the initial steps as presented in the savings algorithm, with the following modifications.

Scope. The scope of the model remains the same as in the previous savings algorithm – to develop a routing plan that minimizes the total distance travelled by:

- delivering to all customers on time and in full;
- not exceeding the vehicle capacity;
- not exceeding the total maximum number of miles allowed to travel for each vehicle;
- delivering to each customer on the list in order of their location on the map. As soon as the starting point has been identified, the routes should be constructed based on delivering ('sweep'ing) to all customers in order.

Constraints. The constraints of this problem are:

- The number of vehicles is limited.
- The total distance travelled is predefined per vehicle per route.
- The total weight to be carried by a vehicle (in our case to total number of cases) is limited.
- The total number of customers to deliver to is defined.

Input data. For the purpose of this example the total number of customers to deliver to is presented in Figure 7.56, where three vehicles are available to transport products to seven customers (C1, C2, ..., C7). The total distance allocated to each vehicle is not to exceed the 100 mile limit with a maximum vehicle capacity of 30 cases. As input data, these parameters can change, where a larger or smaller number of vehicles can be made available for routing, the total number of customers to deliver to can change, where the vehicle capacity can change depending on the vehicle type used. When a transport organization uses a particular type of vehicles, they may not be able to change the input data allocated for the vehicle capacity as well as the total number of miles travelled.

Figure 7.56 Sweep algorithm – input data

No of available vehicles		3	
Maximum allocated distance travel per vehicle/route		100 miles	
Maximum number of cases allocated per vehicle/route		30 cases	
Location	X-coordinate	Y-coordinate	Demand (cases)
Depot	0	0	N/A
C1	6	16	10
C2	18	–25	9
C3	15	17	11
C4	–19	14	18
C5	–17	–28	6
C6	8	–35	12
C7	–12	38	8

Step 1. This step is similar to Step 1 presented in the savings algorithm. This step refers to identifying all customers that placed an order and position these customers on the graph (see Figure 7.57), or in other words, locating the customers on the map using their address.

Step 2. This particular step is identical to Step 2 in the savings algorithm. This requires calculation of the *distance matrix* for any possible combination of paring two customers, as well as between the customers and the depot. For this particular example, it is considered that the forward route is the same as the return route; however, it can be argued that in practice this may not be the case due to road restrictions and variations in speed limitations, or any other issues that may arise during transportation. To calculate the distance matrix, the Pythagoras formula will be used for the distance between two points.

$$d(C1,C2)=\sqrt{(x_{C1}-x_{C2})^2+(y_{C1}-y_{C2})^2}$$

A template is formulated in Excel for calculating the distance matrix, and the formulas to be used in Excel are as follows.

Figure 7.57 VRP sweep algorithm – customers

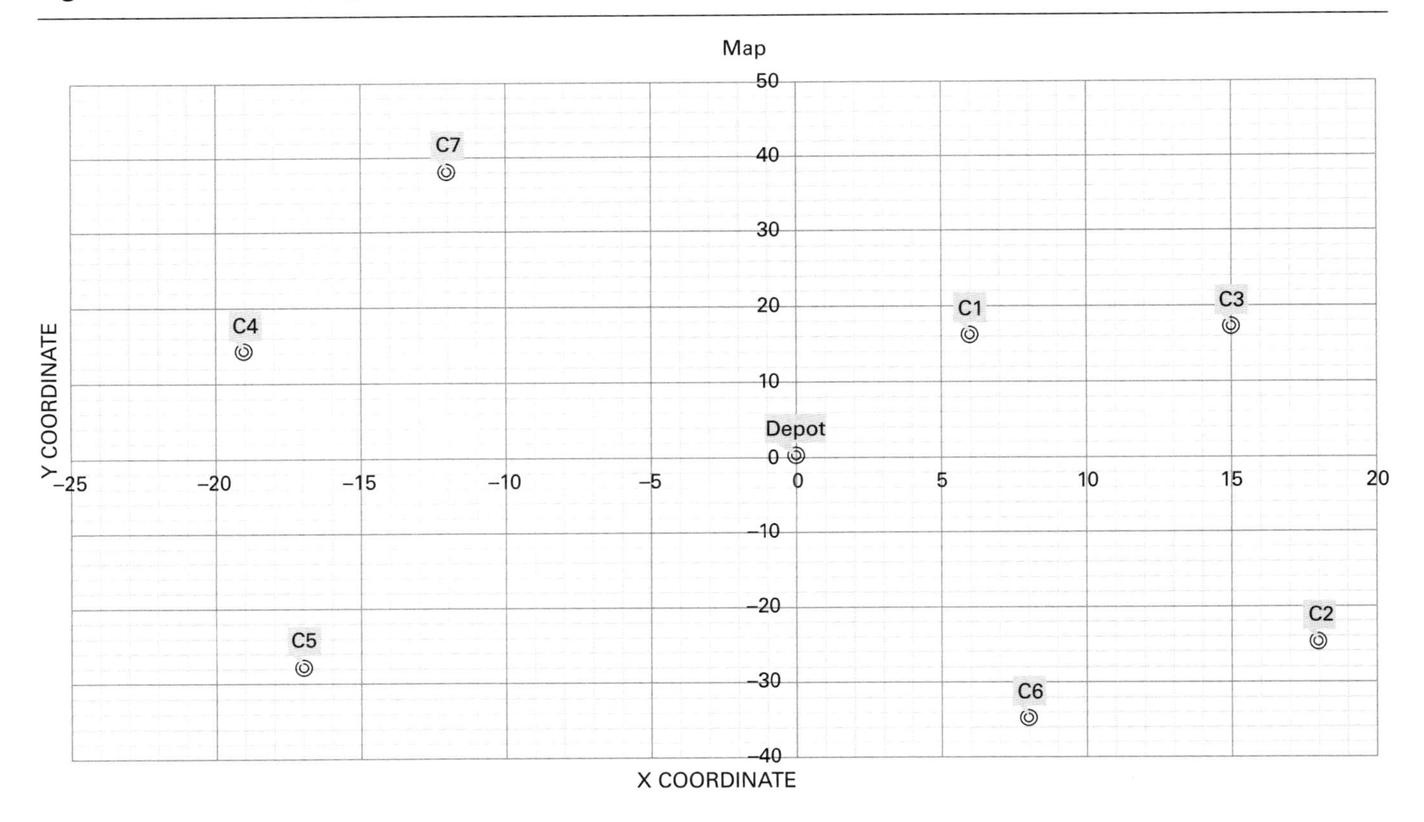

To calculate the distance matrix in Excel, the following formula could be used for any combination for values inserted in the table. Therefore for the distance between the depot and customer 1 we can use: d(D, C1)= =SQRT (SUMXMY2($C$6:$D$6,C7:D7)), where the d(D,C2)= =SQRT(SUMXMY2 ($C$6:$D$6,C8:D8)) and so on.

Figure 7.58 VRP sweep algorithm – distance matrix

Distance Matrix	Depot	C1	C2	C3	C4	C5	C6	C7
Depot	0.00	17.09	30.81	22.67	23.60	32.76	35.90	39.85
C1	17.09	0.00	42.72	9.06	25.08	49.65	51.04	28.43
C2	30.81	42.72	0.00	42.11	53.76	35.13	14.14	69.78
C3	22.67	9.06	42.11	0.00	34.13	55.22	52.47	34.21
C4	23.60	25.08	53.76	34.13	0.00	42.05	55.95	25.00
C5	32.76	49.65	35.13	55.22	42.05	0.00	25.96	66.19
C6	35.90	51.04	14.14	52.47	55.95	25.96	0.00	75.69
C7	39.85	28.43	69.78	34.21	25.00	66.19	75.69	0.00

Step 3. Construct the first routing plan. Within the sweep algorithm, the key elements is to identify the starting point that potentially will generate the shortest distance travelled and meet the constraints of the problem.

If, for example, the starting point considered is to start the routing with C1, then to sweep in a clockwise direction we will have the following:

- Route 1: D–C1–C3–C2–D, with a total distance travelled of d(Route 1-S1) = 90.92 miles and carrying c(Route 1-S1) = 30 cases in total.
- Route 2: D–C6–C5–D, with a total distance travelled of d(Route 2-S1) = 94.62 miles and carrying only c(Route 1-S1) = 18 cases in total. This will result in the vehicle being used to only 60 per cent of its total capacity. However, as the total distance to be travelled on this route is very close to the maximum distance allowed, it will not be possible to include a delivery for any other customer.
- Route 3 is then formed from the remaining two customers, such as: D–C4–C7–D, with a total distance of d(Route 3-S1) = 88.45 miles and a total of c(Route 1-S1) = 26 cases in total.

Figure 7.59 VRP sweep algorithm – solution 1

Plan 1	Route details	Distance travelled per route	Total no of cases	Vehicle capacity used
Route 1	D-C1-C3-C2-D	90.92	30	100%
Route 2	D-C6-C5-D	94.62	18	60%
Route 3	D-C4-C7-D	88.45	26	87%
	Total distance travelled	273.99		

Step 4. Repeat Step 3 for a number of combinations of different starting points and calculate the overall total distance. Compare this distance with the previously identified solution and decide on the best option.

In this particular case, three different starting points have been considered as presented in Figure 7.60 with C1, C2 and C3 as starting points.

In this particular example and from the three examples calculated, it appears that the best solution when looking to minimize the total distance is obtained for Solution 1, therefore Plan 1 with starting point C1. However, Plan 2 gives a better solution if concerned with vehicle utilization, and a very small increase in the total distance travelled. Plan 2 may present a better solution if managers are looking for maintaining a balance in utilizing their vehicles. The solution presented in Plan 3, however, considers a more balanced approach for utilizing two of the vehicles, but it comes with an increase in the total distance travelled.

This example is particularly relevant when considering small deliveries to close locations, such as the 'milk run' or postal deliveries. This application is seen in door-to-door deliveries or last mile delivery examples.

Figure 7.60 VRP sweep algorithm – solutions 1, 2 and 3

Plan 1	Route details	Distance travelled per route	Total no of cases	Vehicle capacity used
Route 1	D-C1-C3-C2-D	90.92	30	100%
Route 2	D-C6-C5-D	94.62	18	60%
Route 3	D-C4-C7-D	88.45	26	87%
	Total distance travelled	273.99		

Plan 2	Route details	Distance travelled per route	Total no of cases	Vehicle capacity used
Route 1	D-C2-C6-D	80.85	21	70%
Route 2	D-C5-C4-D	98.41	24	80%
Route 3	D-C7-C1-C3-D	100.00	29	97%
	Total distance travelled	279.26		

Plan 2	Route details	Distance travelled per route	Total no of cases	Vehicle capacity used
Route 1	D-C3-C2-D	95.58	20	67%
Route 2	D-C6-C5-D	94.62	18	60%
Route 3	D-C4-C7-C1-D	94.11	29	97%
	Total distance travelled	284.32		

Figure 7.61 VRP sweep algorithm – map solution 1

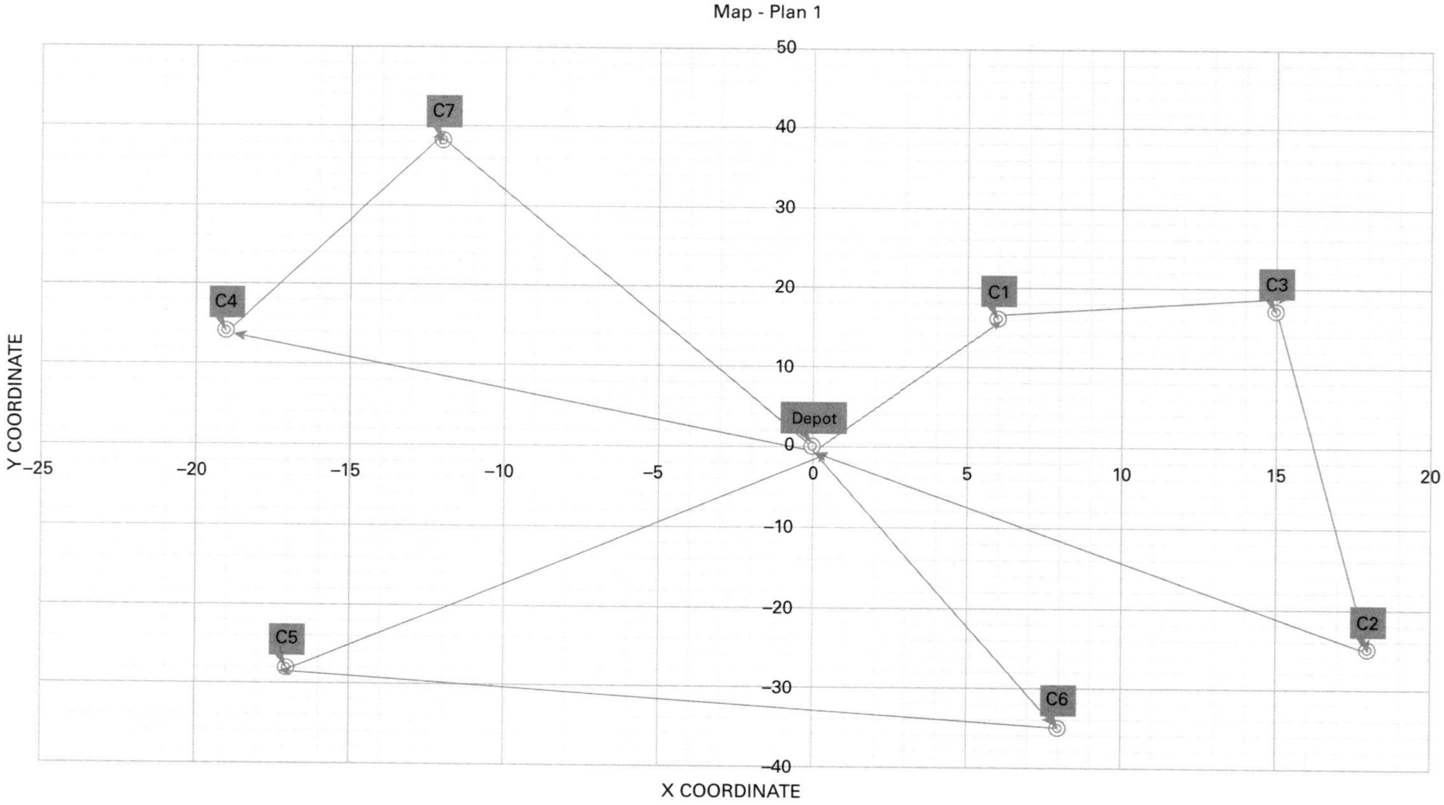

Map - Plan 1

Y COORDINATE

X COORDINATE

Depot

C1

C2

C3

C4

C5

C6

C7

Figure 7.62 VRP sweep algorithm – map solution 2

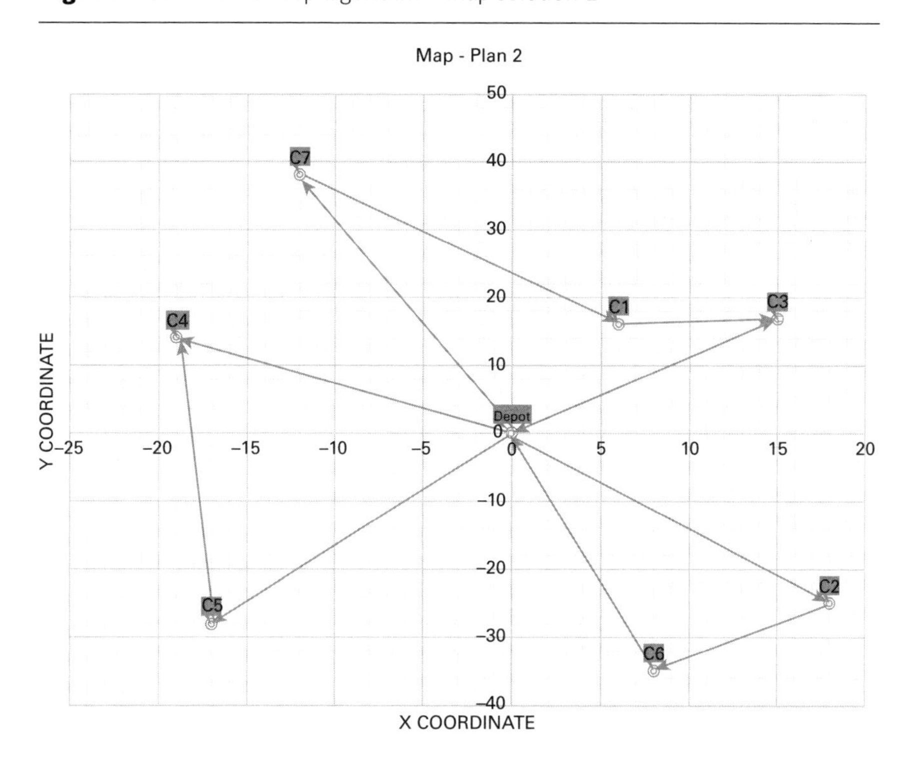

Figure 7.63 VRP sweep algorithm – map solution 3

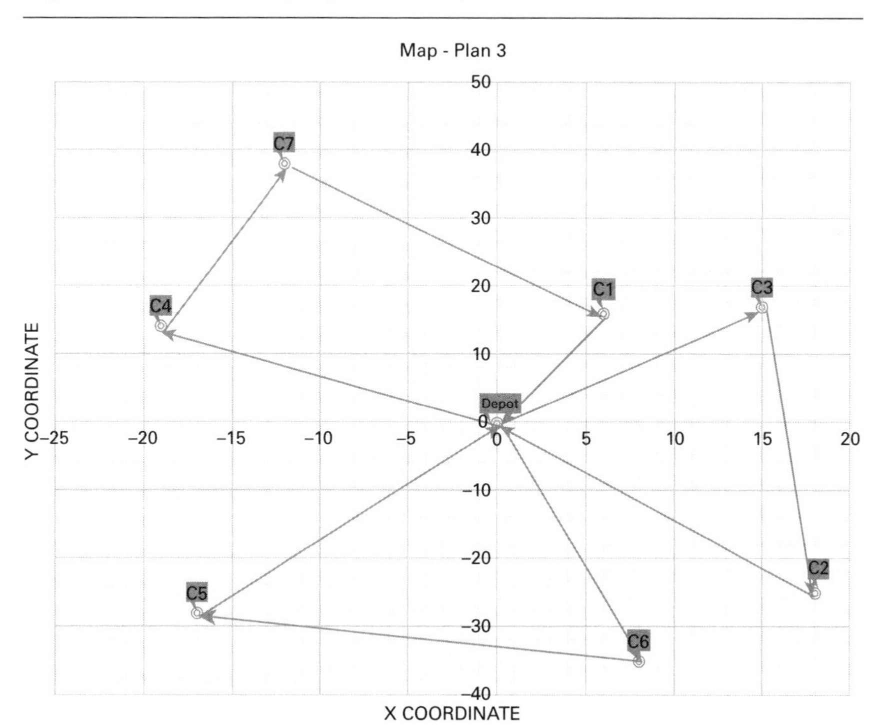

Observations: Both examples (the savings and sweep algorithms) have a number of similar points with regard to considerations given to the constraints to be taken into consideration, such as location, vehicle capacity, distance travelled, customer demand, calculating the distance matrix and setting different scenarios to identify the best overall solution. These can both be used in this format when the number of customers considered is not very large. However, when the number of customers increases, a dedicated VRP software package such as Paragon vehicle routing and scheduling software (https://www.paragonrouting.com/en-gb/our-products/routing-and-scheduling/) (archived at https://perma.cc/B82N-2LBN) can be used.

Summary

A number of models have been presented in this chapter that not only look at providing a solution but also explain how this solution could help the decision-making process in making operational or strategic changes.

The models selected and presented are significant in the field of supply chain. However, the list of models that could be considered in the field of supply chain analytics can be extended. It is interesting to note the particular area within the supply chain network that each of these models can tackle. There is no one single model that could cover the entire set of issues that are present within a supply chain environment. Some of the models only refer to resource allocation and will be relevant to apply in many organizations within the supply chain system, and they can have a holistic view on a number of activities that are taking place within an organization and their partners forming the supply chain. The covering model has applications in any organization that forms part of the supply chain and this can be used when considering daily activities as well as weekly, monthly or yearly activities. The assignment model forms part of strategic decision-making and again is relevant to organizations as part of a larger supply chain. The transportation and transshipment models are particularly applied at the distribution end of the supply chain, from suppliers to various partners, manufacturers and so on, as well as from distributors to different retailers and end customers. The vehicle routing examples used different analytical approach than the examples presented above, but they solve another aspect of issues in the distribution chain. They can be used alone or in combination with the transportation models to answer the delivery questions and ensure the best route for delivery. Individual delivery plans can be obtained in this case with particular reference to the large number of constrains that form part of the distribution network and the type of issues a VRP algorithm can solve. Facility location models, again, are different in the modelling approach, and require a different algorithm, but they solve the allocation of the best facility location based on a number of constraints. This is a model that is used in strategic decisions, when finding the best locations for a facility is required.

Steps in using these models have been given, and the majority of the models used in this section can be solved using spreadsheets and Microsoft Excel add-in Solver.

For each model used, particular stress has been placed on ensuring the input data is available or collected before an analyst can start the modelling process. The output data expected after running the model was also discussed and this is particularly relevant for managers in charge of taking decisions based on the results obtained and implementing change for operation, production and supply chain improvement. Operational performance measures can be easily identified based on the variables used and the objective function set, but not only that. Other performance measures, such as customer service, delivery in full and others can be set when using these models and, in many cases, these are being missed from analysis. Various performance measures have been referred to for each model considered in this chapter.

There are many other models that can be used in the field of analytics and supply chain modelling that are not detailed in this chapter. References to other models already published in the area of operational research and management science that are relevant to the field of logistics and supply chain are provided.

References

Albright, S C and Winston, W L (2005) *Spreadsheet Modelling and Applications: Essentials of practical management science*, South-Western Cengage Learning, Mason, OH

Albright, S C and Winston, W L (2011) *Management Science Modeling*, 4th edn, South-Western College Pub, Andover and Mason, OH

Baker, K R (2016) *Optimization Modeling with Spreadsheets*, 3rd edn, Wiley, Hoboken, NJ

Balakrishnan, N (2017) *Managerial Decision Modeling: Business analytics with spreadsheets*, 4th edn, DeG Press, Boston, MA

Clarke, G and Wright, J W (1964) Scheduling of vehicles from a central depot to a number of delivery points, *Operations Research*, 12(4), 568–81, doi:10.1287/opre.12.4.568

Dantzig, G B and Ramser, J H (1959) The truck dispatching problem, *Management Science*, 6(1), 80–91, doi:10.1287/mnsc.6.1.80

Evans, J R (2016) *Business Analytics: Methods, models, and decisions*, 2nd edn, Pearson, Boston, MA

Laporte, G (2009) Fifty years of vehicle routing, *Transportation Science*, 43(4), 408–16, doi:10.1287/trsc.1090.0301

Lepenioti, K, Bousdekis, A, Apostolou, D and Mentzas, G (2020) Prescriptive analytics: Literature review and research challenges, *International Journal of Information Management*, 50, 57–70, doi:10.1016/j.ijinfomgt.2019.04.003

Powell, S G and Baker, K R (2011) *Management Science: The art of modeling with spreadsheets*, 3rd edn, John Wiley & Sons, Hoboken, NJ

Ragsdale, C T (2018) *Spreadsheet Modeling and Decision Analysis: A practical introduction to business analytics*, 8th edn, Cengage Learning, Boston, MA

Russell, R S and Taylor, B W, III (2014) *Operations and Supply Chain Management*, 8th edn, John Wiley & Sons, Singapore

Taylor, B W, III (2016) *Introduction to Management Science*, 12th edn, Pearson, Boston, MA

Bibliography

Dantzig, G B (2002) Linear programming, *Operations Research*, 50(1), 42–47

Waters, C D J (2011) *Quantitative Methods for Business*, 5th edn, Financial Times/Prentice Hall, Harlow

PART THREE
Future opportunities in supply chain analytics and modelling

The future research agenda of supply chain analytics 08

LEARNING OBJECTIVES

- Understand how systematic reviews of the literature have been conducted in the field of supply chain analytics.
- Evaluate the process used by researchers for conducting systematic literature reviews.
- Discuss the research agendas set by researchers in the field of supply chain analytics that suggest future research directions.

Introduction

This chapter reviews a selection of current studies that have been conducted using a systematic literature review (SLR) methodology aimed at setting the research agenda in the area of supply chain analytics. It also explores what systematic literature reviews are, and how different authors have approached their SLR in the field of supply chain analytics. Therefore, a review of SLRs forms part of the discussion in this chapter, where the future research agenda of this field has been debated. This chapter takes a critical angle in its concluding remarks and puts forward a set of issues and challenges that face this area of research.

Conducting a systematic literature review

A number of researchers have conducted SRLs and there are many established procedures on how to conduct them. Examples of these are provided in Bryman and Bell (2015); Quinlan and Zikmund (2011); Adams et al (2014); Saunders et al (2016) and O'Gorman and Macintosh (2014).

The scope of the discussion in this chapter is not to go into detail about why reviews of the literature are important; instead it seeks to understand the research agenda proposed by different authors who are researching in the field of supply chain analytics after they have conducted a review of the literature and summarised their understanding of the material they have reviewed. This summary is particularly valuable as it shapes our view on what the future of the supply chain analytics research agenda will be.

There are a number of different types of literature reviews that can be carried out and, as O'Gorman and Macintosh (2014) indicate, they are:

- **Traditional or narrative reviews** that are directly intended to answer the research question proposed and, in some cases, propose a theoretical or a conceptual framework.
- **Systematic literature reviews**, which are still intended to answer a research or a set of research questions but are conducted in a systematic way. The methodology established in conducting these reviews is clearly presented and can be replicated by other researchers. A few more details of this type of review will be presented later in this chapter.
- **Meta-analysis reviews of the literature**, where the findings from the literature are considered and established statistical procedures are used.
- **Meta-synthesis literature reviews**, where the analysis is not directly based on statistical methods, but more on the use of qualitative techniques to understand the findings of the literature.

Systematic literature reviews have been used by researchers in the field of supply chain analytics and therefore this approach has been observed and related in this chapter. Supply chain analytics is seen as a new area of research, and authors have considered a systematic review of published material to be able to gather specific details about this field, to identify what are the key themes currently being investigated by researchers and what is known and has been summarized so far. Authors such as Webster and Watson (2002), Tranfield et al (2003) and Denyer and Tranfield (2009) have detailed procedures on what is and how to construct a review of the literature. Denyer and Tranfield (2009) propose a five-steps methodology in conducting a review: (1) formulating the research question; (2) identifying the relevant studies that will form part of the review; (3) selecting and evaluating the studies; (4) analysis and systemic stage; and (5) presentation of the results and final discussion.

1. Formulating the research question (RQ)

Following the steps above, the research question proposed in this case is:

> RQ: *What are the key domains that are taking further the field of supply chain analytics?*

2. Identification of relevant studies

a. Search engine identification

To identify the key studies that will form part of the systematic review, a number of search engines can be used such as SCOPUS, Emerald Journals, ScienceDirect, Web of Science, ABI/INFORMS Complete, EBSCOhost and others. For systematic reviews, where it is essential to capture all the relevant studies that could be incorporated within an analysis, a combination of these platforms should be used.

For example, SCOPUS has been used to present the material in this chapter.

b. Keywords selection

After the search engines have been identified and a justification provided for their selection, the keywords used to carry out the review are to be established. There are procedures to identify the keywords to be used; some could be analytical using the keywords of previously identified articles, or from testing with a team of experts. All relevant keywords should be considered that have the potential to result in selecting relevant articles for the review.

The keywords used in this particular search were: ‘supply chain’ AND ‘analytics’ AND ‘systematic literature reviews’.

c. Restrictions

In SCOPUS the initial restrictions are as follow:

> (TITLE-ABS-KEY (’Supply Chain’)) AND ((analytics)) AND (’systematic literature review’) AND (LIMIT-TO (PUBSTAGE , ’final’)) AND (LIMIT-TO (DOCTYPE , ’ar’)) AND (LIMIT-TO (LANGUAGE , ’English’)) AND (LIMIT-TO (SRCTYPE , ’j’)).

The search is carried out on the article title, abstract and keywords, the publication stage of the article must be final, and the article must be written in English and published in a journal. Therefore, no books, theses, dissertations or articles published in conference proceedings or newspapers have been incorporated in this search.

This initial search returned *266 articles* and the final search was carried out on 9 October 2020. The first set of articles in this group was published in 2015, with no articles being identified before this year.

3. Selecting and evaluating the studies

At this stage, the articles's abstract, name, author name(s), year of publication, journal title and author keywords are extracted and exported into Excel. An analysis table can be developed with the headlines *Authors*, article *Title*, *Year* of publication, *Journal title* (*Source title*), article *Abstract*, *Author keywords* and *Inclusion* or *Exclusion* criteria selection. An example of this type of table can be seen in Figure 8.1. At this stage, researchers read and evaluate the abstract to ensure that the article is in the area they have selected. In our example, the articles should be in the area of supply chain analytics where a systematic literature review has been carried out. When there is more than one researcher involved in carrying out the same systematic literature review, at this point, it is relevant to cross-examine the results from this stage conducted by all researchers to ensure validity of the final list of articles selected. When there are a large number of articles, researchers may decide to select a 10 per cent sample from the articles evaluated by the other researchers in the team. In the case of a discrepancy in the articles selected, the authors will have the opportunity to discuss the results and reinforce the rules selected for the evaluation.

At the end of this stage, reviewing the abstract and the keywords, a total of *41 articles* have been selected for further analysis. These articles were published in the journals listed in Figure 8.2. A few journals have considered two review articles in the past five years, but the majority of identified journals only have one review article in this case, so there is evidently scope for further reviews to be considered in the field of supply chain analytics.

Following this stage, the review went even closer, looking at SLR that deal with issues of analytics in the supply chain, but not if analytics is used in the analysis process for articles considering supply chain issues.

This further restricted the list to *20 articles* and these are listed in Figure 8.3.

Figure 8.1 Example of selected articles inserted in Excel from Scopus

	A	B	C	D	E	F	G	H
1	Authors	Title	Year	Source title	Abstract	Author Keywords	Include	
2	Kouhizadeh M., Saberi S., Sarkis J.	Blockchain technology and the sustainable supply chain: Theoretically exploring adoption barriers	2021	International Journal of Production Economics	Blockchain technology has gained global attention with potential to revolutionize supply chain management and sustainability achievements. The few applied ongoing use cases include blockchain for food, healthcare, and logistics supply chains have emphasized blockchain's untapped potential. Potential support for supply chain and sustainability issues include improving efficiency, transparency, and traceability in addition to billions of dollars in corporate financial savings. Given its promise, the adoption of blockchain technology, although hyped for years, has not seen rapid acceptance. In this study, the technology-organization-environment framework and force field theories are utilized to investigate blockchain adoption barriers. Using various literature streams on technology, organizational practices, and sustainability, a comprehensive overview of barriers for adopting blockchain technology to manage sustainable supply chains is provided. The barriers are explored using technology, organizational, and environmental - supply chain and external - framework followed by inputs from academics and industry experts and then analyzed using the Decision-Making Trial and Evaluation Laboratory (DEMATEL) tool. The results show that supply chain and technological barriers are the most critical barriers among both academics and industry experts. We further determine the similarities and differences among academics and practitioners in perceiving the barriers. This exploratory study reveals interesting relative importance and interrelationships of barriers which are necessary, theoretically and practically for further adoption and dissemination of blockchain technology in a sustainable supply chain environment. It also sets the stage for theoretical observations for understanding blockchain technology implementation in sustainable supply chains. A series of research propositions and research directions culminate from this exploratory study. © 2020 Elsevier B.V.	Barrier analysis; Blockchain; DEMATEL; Supply chain management; Sustainability; Technology-organization-environment framework	Y	
3	Toorajipour R., Sohrabpour V., Nazarpour A., Oghazi P., Fischl M.	Artificial intelligence in supply chain management: A systematic literature review	2021	Journal of Business Research	This paper seeks to identify the contributions of artificial intelligence (AI) to supply chain management (SCM) through a systematic review of the existing literature. To address the current scientific gap of AI in SCM, this study aimed to determine the current and potential AI techniques that can enhance both the study and practice of SCM. Gaps in the literature that need to be addressed through scientific research were also identified. More specifically, the following four aspects were covered: (1) the most prevalent AI techniques in SCM; (2) the potential AI techniques for employment in SCM; (3) the current AI-improved SCM subfields; and (4) the subfields that have high potential to be enhanced by AI. A specific set of inclusion and exclusion criteria are used to identify and examine papers from four SCM fields: logistics, marketing, supply chain and production. This paper provides insights through systematic analysis and synthesis. © 2020 The Author(s)	Artificial intelligence; Supply chain management; Systematic literature review	Y	
4	Cai Y.-J., Lo C.K.Y.	Omni-channel management in the new retailing era: A systematic review and future research agenda	2020	International Journal of Production Economics	Omni-channel retailing is a popular strategy in a new retailing era when digitalization, social media, big data and other emerging technologies (e.g., Artificial Intelligence (AI), virtual reality (VR), augmented reality (AR), blockchain, etc.) are transforming the retail business models. Meanwhile, omni-channel related operations impose a challenge on either well-established firms or new setups that they have to make "right" decisions to fit in the new retail environment. This review paper endeavors to reveal the established knowledge behind the omni-channel retailing literature, generate managerial implications for firms, and provides a guideline for future research. We conduct this systematic review by adopting citation network analysis (CNA). The CNA helps identify seven independent and interdependent research domains, which depict (or constitute) a whole picture of "omnichannel management". The main path analysis reveals that each identified research domain is under study. We also find that the extant literature seldom examines the roles of how new technologies play in the "omnichannel management". Moreover, the domain of supply chain management and inventory management in the omnichannel environment is absent in this systematic literature review. Therefore, we propose prescribed framework for "omnichannel management" (PFOM), which contributes to the literature on "omnichannel management" and provides important managerial applications to the retail firms that plan to implement the omnichannel strategy. © 2020	Omni-channel management; Omni-channel retailing; Omni-channel strategy; Review	Y	
5	Shashi, Centobelli P., Cerchione R., Ertz M.	Agile supply chain management where did it come from and where will it go in the era of digital transformation?	2020	Industrial Marketing Management	In today's dynamic business environment, agile supply chain (ASC) has become a key strategic move to cope with market instability, handle competitive pressures and strengthen operational and organizational performance. Meanwhile ASC is a good example of a strategy drawing heavily on digitization since as a demand chain management it was information-centric and technology-centric from its inception. Yet, despite this relationship, a lack of coherence and clarity around the input of technology for ASC has impeded to portray accurately the relative importance of digitization in ASC strategies. This study provides a comprehensive and integrative review of 90 articles on ASC. By so doing, we contribute to the discussion about digitization in the supply chain in several ways. First, the paper reports descriptively and analytically how technology was addressed within the ASC literature. Second, it maps a nomological network of ASC research. Third, it finds that technology appears as a necessary but not-sufficient enabling factor for ASC deployment. Finally, a research agenda is proposed to suggest future research avenues to improve contributions to ASC performance. © 2020 Elsevier Inc.	Agility measurement; Barriers; Digitalization; Factors; Industry 4.0; Resilience; Structured literature review; Supply chain agility	Y	
6	Nakano M., Lau A.K.W.	A systematic review on supply chain risk management: using the strategy-structure-process-performance framework	2020	International Journal of Logistics Research and Applications	The purpose of this study is to conduct a systematic literature review on the supply chain risk management (SCRM) in the concepts of strategy, structure, process, and performance (SSPP). Using the EBSCO, Science Direct and SCOPUS databases and a combination of SCRM keywords, 2851 articles were identified from the journals in year 2001–2017. Using the SSPP framework, 174 articles were reviewed as the basis for the analysis. This study firstly fits a matrix of 'redundant-flexible' organisational design with internal and external organisational structures and processes. Using this framework, supply chain scholars and managers may discuss the organisational fit of SCRM operations comprehensively. It also identifies the new areas for further studies. © 2019 Informa UK Limited, trading as Taylor & Francis Group.	strategy-structure-process-performance framework; Supply chain risk management; systematic literature review	Y	
7	Le P.L., Elmughrabi W., Dao T.-M., Chaabane A.	Present focuses and future directions of decision-making in construction supply chain management: a systematic review	2020	International Journal of Construction Management	This paper utilizes a systematic literature review methodology to identify the present focuses and discuss the future directions of decision-making in construction supply chain management (CSCM). The results show that, at present, the CSCM applications are still focusing on material and resources management with the internal supply chain (SC) integration. Strategic decisions related to building partnerships, IT-based planning and logistics-based planning are not conducted at the early phase of planning and design. For future trends in CSCM application, a framework is proposed to leverage the three important points: the collaborative planning and design with advanced techniques; the lean procurement with BIM and third-party logistics; and the application of Lean and BIM in construction operations and delivery. The original contribution of this paper is the attempt of identifying CSC decisions and suggests how they should be delivered during the phases of a construction project with the use of appropriate SC methods and tools. © 2018, © 2018 Informa UK Limited, trading as Taylor & Francis Group.	Construction supply chain management; decision-making; future directions; present focuses	Y	
	Kumar M.S., Raut D.R.D., Narwane	Applications of industry 4.0		Diabetes and Metabolic Syndrome: Clinical	Background and aims: An epidemic outbreak of COVID-19 has increased the demand for medical equipment, medical accessories along with daily essentials for the safety of healthcare workers. This study aims to identify the operational challenges faced by retailers in providing effiecient services. The study also aimed to propose the roadmap of Industry 4.0 to reduce the impact of COVID-19. Methods: A detailed literature review is done on an epidemic outbreak and supply chain using appropriate keywords on SCOPUS, Science Direct, Google Scholar. Some relevant industry reports and blogs are also taken to get insights. Results: We have identified twelve significant challenges for the retail sectors that are acting as operational barriers and provided the application of Industry 4.0 technologies to deal with it. Conclusion: Industry 4.0 can act as a significant driver for reducing the impact of identified challenges on retailers to fight against the pandemic. There is a need to build trust and transparency for the effective management of healthcare essentials. The supply chain partners and government bodies should act wisely for improving.	Coronavirus; COVID-19; Epidemic outbreak;		

Figure 8.2 Round 2 selected journals for the SLR

Journals	No of articles
Computers and Industrial Engineering	2
Computers and Operations Research	1
Diabetes and Metabolic Syndrome: Clinical Research and Reviews	1
EMJ - Engineering Management Journal	1
European Journal of Operational Research	1
Global Journal of Flexible Systems Management	2
ICIC Express Letters	1
IEEE Access	1
IEEE Engineering Management Review	1
Industrial Marketing Management	1
International Journal of Advanced Operations Management	1
International Journal of Construction Management	1
International Journal of Engineering Research and Technology	1
International Journal of Integrated Supply Management	1
International Journal of Logistics Management	2
International Journal of Logistics Research and Applications	2
International Journal of Logistics Systems and Management	1
International Journal of Management and Decision Making	1
International Journal of Manufacturing Technology and Management	1
International Journal of Physical Distribution and Logistics Management	1
International Journal of Production Economics	2
International Journal of Production Research	2
International Journal of Supply Chain Management	1
Journal of Business Research	1
Journal of Cleaner Production	1
Journal of Manufacturing Systems	1
Knowledge Management Research and Practice	1
Management Decision	1
Resources, Conservation and Recycling	1
Science of the Total Environment	1
Supply Chain Management	2
Sustainability (Switzerland)	1
Transportation Research Part E: Logistics and Transportation Review	1
World Review of Intermodal Transportation Research	1
Total	41

Figure 8.3 Articles selected for the SLR

No.	Authors	Year	Title	Journal name
1	Arunachalam D., Kumar N., Kawalek J.P.	2018	Understanding big data analytics capabilities in supply chain management: Unravelling the issues, challenges and implications for practice	Transportation Research Part E: Logistics and Transportation Review
2	Calatayud A., Mangan J., Christopher M.	2019	The self-thinking supply chain	Supply Chain Management
3	Chiappetta Jabbour C.J., Fiorini P.D.C., Ndubisi N.O., Queiroz M.M., Piato É.L.	2020	Digitally-enabled sustainable supply chains in the 21st century: A review and a research agenda	Science of the Total Environment
4	Frederico G.F., Garza-Reyes J.A., Anosike A., Kumar V.	2019	Supply Chain 4.0: concepts, maturity and research agenda	Supply Chain Management
5	Giri C., Jain S., Zeng X., Bruniaux P.	2019	A Detailed Review of Artificial Intelligence Applied in the Fashion and Apparel Industry	IEEE Access
6	Grover P., Kar A.K.	2017	Big Data Analytics: A Review on Theoretical Contributions and Tools Used in Literature	Global Journal of Flexible Systems Management
7	Jede A., Teuteberg F.	2016	Towards cloud-based supply chain processes: Designing a reference model and elements of a research agenda	International Journal of Logistics Management
8	Novais L., Maqueira J.M., Ortiz-Bas Á.	2019	A systematic literature review of cloud computing use in supply chain integration	Computers and Industrial Engineering
9	Núñez-Merino M., Maqueira-Marin J.M., Moyano Fuentes J., Martínez-Jurado P.J.	2020	Information and digital technologies of Industry 4.0 and Lean supply chain management: a systematic literature review	International Journal of Production Research
10	Perera H.N., Hurley J., Fahimnia B., Reisi M.	2019	The human factor in supply chain forecasting: A systematic review	European Journal of Operational Research
11	Pérez-Pérez M., Kocabasoglu-Hillmer C., Serrano-Bedia A.M., López-Fernández M.C.	2019	Manufacturing and Supply Chain Flexibility: Building an Integrative Conceptual Model Through Systematic Literature Review and Bibliometric Analysis	Global Journal of Flexible Systems Management
12	Raut R.D., Gotmare A., Narkhede B.E., Govindarajan U.H., Bokade S.U.	2020	Enabling Technologies for Industry 4.0 Manufacturing and Supply Chain: Concepts, Current Status, and Adoption Challenges	IEEE Engineering Management Reviews
13	Rialti R., Marzi G., Ciappei C., Busso D.	2019	Big data and dynamic capabilities: a bibliometric analysis and systematic literature review	Management Decision
14	Safriyana, Marimin, Anggraeni E., Sailah I.	2020	Spatial-intelligent decision support system for sustainable downstream palm oil based agroindustry within the supply chain network: A systematic literature review and future research	International Journal of Supply Chain Management
15	Sharma R., Kamble S.S., Gunasekaran A., Kumar V., Kumar A.	2020	A systematic literature review on machine learning applications for sustainable agriculture supply chain performance	Computers and Operations Research
16	Shashi, Centobelli P., Cerchione R., Ertz M.	2020	Agile supply chain management: where did it come from and where will it go in the era of digital transformation?	Industrial Marketing Management
17	Surjandy, Meyliana, Hidayanto A.N., Prabowo H.	2019	The latest adoption blockchain technology in supply chain management: A systematic literature review	ICIC Express Letters
18	Toorajipour R., Sohrabpour V., Nazarpour A., Oghazi P., Fischl M.	2021	Artificial intelligence in supply chain management: A systematic literature review	Journal of Business Research
19	Viet N.Q., Behdani B., Bloemhof J.	2018	The value of information in supply chain decisions: A review of the literature and research agenda	Computers and Industrial Engineering
20	Vishnu C.R., Sridharan R., Kumar P.N.R.	2019	Supply chain risk management: Models and methods	International Journal of Management and Decision Making

Figure 8.4 Examples of different SLR approaches

Authors	Year	SLR process used	Papers reviewed	SLR years of selected publications
Arunachalam D., Kumar N., Kawalek J.P.	2018	(1) Literature search, with (1a) Inclusion and exclusion criteria; (2) findings and discussion; (3) thematic analysis	82	2008 to 2016
Calatayud A., Mangan J., Christopher M.	2019	(1) problem formulation; (2) literature research; (3) selection and evaluation of literature; (4) research analysis and interpretation; and (5) presentation of results	126	1950 to 2018
Chiappetta Jabbour C.J., Fiorini P.D.C., Ndubisi N.O., Queiroz M.M., Piato É.L.	2020	(1) Identification and selection of publications; (2) Data extraction from publications; (3) Results and discussion: current lessons and future research avenues	33	2015 to 2019
Frederico G.F., Garza-Reyes J.A., Anosike A., Kumar V.	2019	(1) Planning; (2) conducting the search; (3) reporting	24	2011 to 2018
Grover P., Kar A.K.	2017	(1) Literature search; (2) analysis of literature; (3) presentation of results	118	2007 to 2017
Jede A., Teuteberg F.	2016	(1) Definition of the research field; (2) Literature search; (3) Literature evaluation and articles extraction; (4)a qualitative analysis via predetermined framework; (4)b quantitative analysis; (5) Literature reunification and results presentation; (6) Reference model development and refinement; (7) Reference model evaluation: (8) Future research questions;	102	no restrictions
Novais L., Maqueira J.M., Ortiz-Bas Á.	2019	(1) Formulation of research questions; (2) Identification of study; (3) Selection and evaluation of studies; (4) Analysis and synthesis; (5) Presentation of results.	77	no restrictions
Perera H.N., Hurley J., Fahimnia B., Reisi M.	2019	(1) Initial search; (2) Systematic exclusion of irrelevant papers; (3) Manual pruning of search results; (4) Verification with reference lists; (5) Replicating the search terms on Web of Science; (6) Crosschecking top journals; (7) Inputs from prolific authors	113	1981 to 2017
Pérez-Pérez M., Kocabasoglu-Hillmer C., Serrano-Bedia A.M., López-Fernández M.C.	2019	(1) Collection of past literature and evaluation for appropriatness; (2) Thematic identification through bibliometric analysis: co-word technique; (3) Labelling the thematic clusters and critical reflection	222	1996 to 2018
Raut R.D., Gotmare A., Narkhede B.E., Govindarajan U.H., Bokade S.U.	2020	(1) initiate the review process through a systematic keyword search across public databases; (2) a detailed review of selected literature; and (3) evidencebased summary of findings	86	2015 to 2019
Sharma R., Kamble S.S., Gunasekaran A., Kumar V., Kumar A.	2020	(1) planning the review; (2) conducting the review; (3) descriptive statistics; (4) review, findings and discussion	93	2002 to 2019
Viet N.Q., Behdani B., Bloemhof J.	2018	(1) Question formulation; (2) Locating studies; (3) Study selection and evaluation; (4) Analysis and synthesis; (5) Reporting and using the results	117	2006 to 2017

4. Analysis and systemic stage

This stage continues with the 20 articles. To understand the review process and the different approach authors may select when carrying out an SLR, a sample of articles has been selected and compared in Figure 8.4. This indicates the different approaches used in each article, the total number of articles selected for review and the years from which the articles have been selected.

Some authors review articles starting with 1950 (Calatayud et al, 2019); however, this was based on the specific research question they consider. Other authors look at articles over a short period of time from 2015 to 2019 (Chiappetta Jabbour et al, 2020; Raut et al, 2020). From Figure 8.4 it can be noted that there are differences in the approach used from author to author, and in the total number of articles selected for review.

A selection of *themes identified* from the reviewed articles is captured in Table 8.1.

Table 8.1 Identified themes from the SLR

Data and technologies	Technologies	Industry 4.0	Supply chain
Big data; big data analytics	Data science	Circular economy	Big data-driven sustainable supply chains
Value of big data	Information and digital technologies	Industry 4.0	Supply Chain 4.0
Expert systems	Challenges in technology adoption	Capabilities maturity model	Supply chain risk management
Machine learning	Blockchain adoption in the SC	Smart farming	Self-thinking supply chains
Cloud computing	Enabling technologies	Advanced manufacturing	Sustainable supply chain
Decision support systems	Demand forecasting	Manufacturing flexibility	Supply chain integration
Digitalization	Lifecycle technology	Performance management	Lean supply chain management

(continued)

Table 8.1 (Continued)

Data and technologies	Technologies	Industry 4.0	Supply chain
Artificial Intelligence	ICT developments with autonomous and predictive capabilities	Smart factory	Resilience
Information accuracy, completeness and timeliness	Big data technologies	Dynamic capabilities	
Information systems	New technology		
Internet of Things			

5. Presentation of the results and final discussion

A number of the themes captured in Table 8.1 have been discussed to some extend in the previous chapters of this book. However, other themes are new (smart factory, smart manufacturing) and others, however current and relevant, have not been directly approached in this book (blockchain) or are topics in the field of the supply chain that have been previously covered by many publications (supply chain risk, sustainable supply chain, supply chain integration and lean supply chain).

Some areas that are closer to the content covered in this book are further discussed and summarized from the point of view of the SLRs collected for this investigation.

Based on the themes identified above, we see reference made to big data (Arunachalam et al, 2018; Rialti et al, 2019), big data analytics (BDA) (Arunachalam et al, 2018; Chiappetta Jabbour et al, 2020; Giri et al, 2019; Grover and Kar, 2017), value captured in big data (Arunachalam et al, 2018; Viet et al, 2018) and big data driven sustainable supply chain (Chiappetta et al, 2020).

A number of publications are particularly concerned with the developments in information systems (Calatayud et al, 2019; Jede and Teuteberg, 2016; Núñez-Merino et al, 2020), cloud computing (Jede and Teuteberg, 2016; Novais et al, 2019; Raut et al, 2020), artificial intelligence

(Toorajipour et al, 2021; Giri et al, 2019; Calatayud et al, 2019), Internet of Things (Calatayud et al, 2019; Raut et al, 2020), blockchain (Surjandy et al, 2019; Raut et al, 2020) and machine learning (Sharma et al, 2020).

A number of very interesting points have been captured in Frederico et al (2019) regarding Supply Chain 4.0 that incorporates the concepts of Industry 4.0 and the number of technologies that could be considered in this regard. A supply chain maturity framework is proposed here, aiming to present a strategy map taking us through different processes at different stages, from the initial stage to the intermediate, advanced and cutting-edge stages. A number of technologies have been identified as disruptive technologies (21, including, among others, robotics, nanotechnology, augmented reality, 3D printing, machine to machine, and others already mentioned in this section) as well as process performance requirements, strategic outcomes and the managerial and capability support required.

Researchers conclude that this area of investigation is still considered new and the full range of benefits and their implementation and post-implementation challenges are still to be further understood.

Taking a closer look at just some of the research agendas, the discussion of each point presented is further emphasized and challenged below in the context of supply chain analytics and modelling. In the majority of cases we note that a need has been identified by all authors to further investigate this area of research with more practical case studies to bring further details in understanding the real challenges and benefits of adopting these technologies and engaging in training staff to accept a data-driven culture as well as gain the required skills to easily manipulate these tools and techniques.

Exploring research agendas proposed in current SLRs

Arunachalam et al (2018) are concerned with the data capability perspective and provide five dimensions: data generation; data integration and management; advanced analytics; data visualization; and data driven culture. They also refer to the importance of cloud computing and absorptive capacity as additional themes to those identified in relation to data capability. It is also interesting to note that their four top journals where their reviewed papers were published were in *International Journal of Operational Research*, *Annals of Operations Research*, *International Journal of Production Economics* and *Decision Support Systems*, although their review article is published in *Transportation Research Part E*. Getting back to the data capabilities, Arunachalam et al (2018) have also developed a BDA capabilities framework for a supply chain, indicating four stages that data and analytics process can take, from a data poor and analytics poor stage moving to data rich and analytics rich.

This study captures aspects of data generation and discusses how organizations in the supply chain should seek the opportunity of generating rich quality data as well as acquiring data from different heterogeneous data sources. It is also acknowledged here that data shared with other partners in the supply chain has the opportunity to increase the supply chain performance. However, in a supply chain there are a number of reasons for reluctance to share data between partners. Aspects are linked to sharing sensitive data, access to quality data by different partners in the chain that is valid and seen as the single point of truth, as well as the fear of losing competitive advantage. Aspects of sharing data in the supply chain could be standardized where partners forming part of the chain could engage in agreeing on different aspects of data that can be share without compromising the sensitivity aspect as well as maintaining organizations' competitiveness. Therefore, Arunachalam et al (2018) propose the consideration of data integration and management capability that that has the ability to collect, integrate and store data from various sources and make this available to organizations in the chain in real time. The importance of advanced analytics and visualization capabilities are also discussed in their study. Within the advanced analytics, aspects of analytics in plan, make, source and deliver are captured together with sustainable supply chain analytics, real-time analytics, in-memory analytics, data mining, web and text mining, trend analytics, and data analysis and data decision capability. It is also very interesting to note the consideration given to data-driven culture not only from an organization point of view but also from a supply chain perspective. Categories captured here include culture and execution, culture and political issues, culture capability, people, culture competence and data and analytics as an asset for decision-making.

As directions for further research in this filed, Arunachalam et al (2018) present the following points that are being further discussed:

- Research should consider the development of models and tools to analyse and integrate other forms of data, such as data captured in video, image data, sensors data, GPS and others forms not fully explored. To further develop the point from Arunachalam et al (2018), it is relevant to indicate that it is not only the generation of data as it is the storage and creating devices that allow for maintenance of data over a longer period of time. Data generated and saved in a video format from 20 years ago is no longer easily accessible in the original format, due to updates in technology. New technology working with different data types accepts electronic files in formats from previous versions, but not files that have been saved, for example, on video tapes, cassettes, photo films and others. Do we fully understand the issues linked to data longevity? Do we have sufficient knowledge on the methodologies of collecting and analysing data that is not in the usual formats we are accustom to?

- Data quality could be an issue if not carefully checked for accuracy, continuously monitored and store appropriately. As noted earlier on in Chapter 4, data quality could greatly affect the supply chain performance. Attention should be given to research into tools and techniques that continuously monitors for data quality. This point, however, requires a number of resources to be put in place as well as reliable systems where partners in the supply chain need to be responsible for taking ownership in managing the data quality.
- Further investigations should be considered regarding in-memory analytics in the supply chain as well as data-driven decision-making tools. There is an indication in Arunachalam et al's (2018) review that we have seen a paradigm shift from heuristic to data-driven models. However, from research into modelling supply chain systems (see Chapter 2 in this book) we still have limited models that capture the complex nature and characteristics of the supply chain. Therefore, the argument that new heuristic models give further understanding to the field of supply chain and supply chain analytics needs further investigations.
- Big data analytics (BDA) has been successfully implemented and there are benefits already reported in the literature. Still, to further understand the positive benefits of the use and implementation of BDA, Arunachalam et al (2018) confirm that further study is needed, as well as applications to more business cases in order to understand the complex nature of BDA implications in practice.
- It has been observed that the implementation of BDA could generate system changes, therefore empirical studies that capture the implementation challenges and benefits of BDA in the context of supply chain and more specifically supply chain networks are still expected. Chapters 2 and 5 of this book have captures aspects of supply chain system structure and design; as new technologies have an effect on changing the structure of the supply chain, further analyses are expected in this regard.

- Arunachalam et al (2018) also ask what the benefits of BDA implementation are to small and medium-size companies, a point also noted by Raut et al (2020). As BDA require the new level of technology, skills and applications, they are more accessible to larger organizations, but there is not only the 'nice to have' technology, there is also a need to have and be able to access the new advances in technology.

Following the issue of big data for sustainable supply chains, Chiappetta et al (2020) have identified gaps in the literature that indicate issues for further investigation in the data-driven sustainable supply chains and give recommendations to practitioners of what should be considered when using big data for sustainable supply chains. The studies selected for their review have been published in journals such as the *Journal of Cleaner Production*, *Sustainability*, *Computers and Operations Research*, *International Journal of Production Economics*, *Resources Conservation and Recycling* and *Technological Forecasting and Social Change*, and their study is published in *Science of the Total Environment*.

One of the gaps indicated in their research is to use big data to model supply chains for sustainability. A number of models and modelling techniques from the area of business analytics have been used to model sustainable supply chains for many years. If big data had been incorporated in these models and in their modelling process more details could have been revealed; however, it is evident that the technology may allow for modelling with very large volumes of data, though not necessarily different formats of data. Modelling with multiple data formats will require more than one analytical tool to be considered for investigation. This is to say that simulation models have been used for modelling sustainable supply chain for many years and many of these models can cope with a large volume of data (in general in numerical format). However, there is a need for more reported examples in the literature that cover the aspect of incorporating more big data in modelling sustainable supply chain systems, as well as supply chain systems in general.

- One other identified gap reported here is the exploration of the big data interface with other Industry 4.0 technologies for promoting sustainable supply chains. This is a point not only expected when working with sustainability in the supply chain, but also with any data-related and process implementation aspects as part of the supply chain activities and links when integrating their functions.

- As in the study by Arunachalam et al (2018), the aspect of data-driven culture is also captured as a gap in the study from Chiappetta et al (2020).
- The implications of BDA in the use and modelling of closed-loop supply chain is another gap that has been identified. Modelling closed-loop supply chains present a number of challenges in relation to understanding the operations and where the main data is coming from. Adding aspects of big data into this context appears as a need in modelling and analytics research.
- Other gaps related in this work were linked with the challenges companies face when implementing BDA, an aspect that has been see before from work from Arunachalam et al (2018) as well as the challenges discussed in Chapter 1 of this book. Still, these challenges are related in the context of sustainable supply chain management.
- Further empirical investigations are also put forward within this agenda to understand the link between resilience and sustainability in the supply chain from the point of view of big data.
- One other point is understanding an organization's capability in handling challenges as well as promoting big data technologies. These points are present in many other studies reviewed here (Raut et al, 2020).

Viet et al (2018) looked at the value of information in supply chain decisions from four dimensions perspectives: supply chain decisions, information dimension, modelling approach, and context of supply chain parameters dimension. From the point of view of the information dimension, the authors identify nine types of information: demand, supply, inventory, planning, product, manufacturing process, transportation process, information regarding the return of products, and public information. It is interesting to note that these types of information dimensions are link to the physical flow of a product, but what is missing here is information regarding the IT systems, where the information is stored and transferred in the supply chain, and information about the supply chain network and the route the product follows. The information sources and characteristics of information are also discussed in this context. The type of decision support tools is also captured in this study and they are listed as: analytical models, game theory, dynamic programming, integer programming, stochastic programming, mixed-integer programming, heuristics, data mining, big data analytics, forecasting, Monte Carlo simulations, system dynamics and discrete event simulations. The context dimension is captured in relation to the value of information, for example the value of information increases when the level of uncertainty increases. On the supply chain decision dimension, we see aspects of inventory decisions, transportations decisions, sourcing and pricing decisions.

As further research directions, the authors suggest aspects of (1) supply chain decisions that should move beyond inventory decision and look more at interdependency decisions; (2) further attention to be given to information characteristics in the SC; (3) more diverse methods in evaluating the value of information; (4) to look at the dynamic, multi-facet aspect of value of information.

Raut et al (2020) review articles published between 2005 and 2019. The first few journals where these articles have been published are: the *International Journal of Production Research*, *Computer and Industrial Engineering*, *International Journal of Production Economics*, followed by *IEEE Transactions on Industrial Informatics*, *International Journal of Information Management* and *Industrial Management and Data Systems*. Their particular article is published in *IEEE Engineering Management Review*. Their review has captured a number of aspects of the use of RFID, Internet of Things, cloud computing, blockchain and big data analytics technology in manufacturing and supply chain management. For each of these technologies, the authors identify that there are barriers to adoption; however, there are identified benefits in each case. Some of the benefits for the RFID, on top of product tracking, are noted on product quality and access to information in real time that can be visible and shared throughout the supply chain and allows for process automation, leading to increase productivity. The IoT technology highlighted benefits real-time information availability; resource tracking also enhances operation efficiency and customer satisfaction. Similar points have also been identified in the review from Calatayud et al (2019): a large quantity of information collected, stored and shared, and supply chain partners can be quickly informed of critical events. It also had a positive outcome on a case in reverse logistics. In the case of cloud computing adoption, the cases studied indicate benefits for optimized manufacturing resources and capabilities allocation, and improved decision making due to the fact that this allows for access to information from any location in the supply chain. Similar benefits are listed for the adoption of BDA and they are in line with benefits summarized by other authors. In the case of blockchain adoption, it is reported here that not many researchers have investigated the adoption and implementation benefits and this would be one area for further investigation. Still, the review indicates benefits of product traceability, transparency, product safety and reduced recall of products.

Similar aspects have been related here as future research directions for the adoption and implementation of these technologies in the supply chain. Cases have been reported, but they are limited to single echelons of the supply chain. It is proposed that the benefits of new technology to multiple echelons of the supply chain should also be assessed. Similar points have been put forward by other authors (see Arunachalam et al, 2018) on the need to investigate the adoption of new technologies by small and medium-size organizations, and by organizations in developing countries. Chiappetta et al (2020) and Raut et al (2020) also indicate that the adoption of these technologies from a sustainable point of view needs to receive further attention and consideration.

The points highlighted so far are present in many of the other reviews, and there is a clear message that this area of investigation is new and further examples are required, to fully understand the benefits and implications of using analytics and modelling in understanding and improving the performance of the entire supply chain system.

Summary

This chapter has reviewed what other authors are presenting as the future research agenda, based on SLRs for topics in the area of analytics and supply chain modelling. This chapter highlighted a few aspects of how reviews of the literature have been conducted in this area and at the same time the material has been presented as a form of a systematic review of the literature. A set of 20 articles were selected for further investigation and themes selected from these reviews. The research agendas in each case were evaluated and it was observed that there are a number of similar actions indicated by authors as future research issues in the field of supply chain analytics. Many of the points indicated in the previous chapters in this book are also captured in the summary of the reviewed articles. A number of summarized points have not been fully detailed here, as they were outside the scope of this book, although they are noted in part. Areas that have not been detailed in this book are simulation technologies and applications that are now emerging in many applied cases in the industry, with many new developments and cases published in the literature. However, this author has contributed to a systematic review of the literature (Chilmon and Tipi, 2020) in the area of simulation methodologies applied in the supply chain where further details can be noted. Note also that aspects of smart factories, smart farming (see the simulation example presented in Gittins et al, 2020) and blockchain technologies have not been detailed here, although they are aspects that are current and form part of the supply chain analytics agenda.

The ability to create models and analyse supply chain in connections, with the support of reliable and adequate data and the use and implementation of new technologies, provides the opportunity to enhance the overall supply chain performance and further our understanding of supply chain systems.

References

Adams, J, Khan, H T A and Raeside, R (2014) *Research Methods for Business and Social Science Students*, 2nd edn, SAGE Publications, India

Bryman, A and Bell, E (2015) *Business Research Methods*, 4th edn, Oxford University Press, Oxford

Chilmon, B and Tipi, N S (2020, in press) Modelling and simulation considerations for an end-to-end supply chain system, *Computers and Industrial Engineering*, doi:10.1016/j.cie.2020.106870

Denyer, D and Tranfield, D (2009) *Producing a Systematic Review: The SAGE handbook of organizational research methods*, SAGE Publications, London, 671–89

O'Gorman, K and Macintosh, R (2014) *Research Methods for Business and Management: A guide to writing your dissertation*, Goodfellow Publishers Limited, Oxford

Quinlan, C and Zikmund, W G (2011) *Business Research Methods*, Cengage Learning EMEA, Andover

Saunders, M, Lewis, P and Thornhill, A (2016) *Research Methods for Business Students*, 7th edn, Pearson Education, Harlow

Tranfield, D, Denyer, D and Smart, P (2003) Towards a methodology for developing evidence-informed management knowledge by means of systematic review, *British Journal of Management*, 14(3), 207–22, doi:10.1111/1467-8551.00375

Webster, J and Watson, R T (2002) Analyzing the past to prepare for the future: Writing a literature review, *MIS Quarterly*, 26(2), xiii–xxiii

SLR references

Arunachalam, D, Kumar, N and Kawalek, J P (2018) Understanding big data analytics capabilities in supply chain management: Unravelling the issues, challenges and implications for practice, *Transportation Research Part E: Logistics and transportation review*, 114, 416–36, doi:10.1016/j.tre.2017.04.001

Calatayud, A, Mangan, J and Christopher, M (2019) The self-thinking supply chain, *Supply Chain Management*, 24(1), 22–38, doi:10.1108/SCM-03-2018-0136

Chiappetta Jabbour, C J, Fiorini, P D C, Ndubisi, N O, Queiroz, M M and Piato, É L (2020) Digitally-enabled sustainable supply chains in the 21st century: A review and a research agenda, *Science of the Total Environment*, 725, doi:10.1016/j.scitotenv.2020.138177

Frederico, G F, Garza-Reyes, J A, Anosike, A and Kumar, V (2019) Supply chain 4.0: Concepts, maturity and research agenda, *Supply Chain Management*, doi:10.1108/SCM-09-2018-0339

Giri, C, Jain, S, Zeng, X and Bruniaux, P (2019) A detailed review of artificial intelligence applied in the fashion and apparel industry, *IEEE Access*, 7, 95376–96, doi:10.1109/ACCESS.2019.2928979

Gittins, P, McElwee, G and Tipi, N (2020) Discrete event simulation in livestock management, *Journal of Rural Studies*, 78, 387–98, https://doi.org/10.1016/j.jrurstud.2020.06.039 (archived at https://perma.cc/9V6J-9BVR)

Grover, P and Kar, A K (2017) Big data analytics: A review on theoretical contributions and tools used in literature, *Global Journal of Flexible Systems Management*, 18(3), 203–29, doi:10.1007/s40171-017-0159-3

Jede, A and Teuteberg, F (2016) Towards cloud-based supply chain processes: Designing a reference model and elements of a research agenda, *International Journal of Logistics Management*, 27(2), 438–62, doi:10.1108/IJLM-09-2014-0139

Kashav, S, Centobelli, P, Cerchione, R and Ertz, M (2020) Agile supply chain management: Where did it come from and where will it go in the era of digital transformation? *Industrial Marketing Management*, 90, 324–45, doi:10.1016/j.indmarman.2020.07.011

Novais, L, Maqueira, J M and Ortiz-Bas, Á (2019) A systematic literature review of cloud computing use in supply chain integration, *Computers and Industrial Engineering*, 129, 296–314, doi:10.1016/j.cie.2019.01.056

Núñez-Merino, M, Maqueira-Marín, J M, Moyano-Fuentes, J and Martínez-Jurado, P J (2020) Information and digital technologies of Industry 4.0 and Lean supply chain management: A systematic literature review, *International Journal of Production Research*, 58(16), 5034–61, doi:10.1080/00207543.2020.1743896

Perera, H N, Hurley, J, Fahimnia, B and Reisi, M (2019) The human factor in supply chain forecasting: A systematic review, *European Journal of Operational Research*, 274(2), 574–600, doi:10.1016/j.ejor.2018.10.028

Pérez-Pérez, M, Kocabasoglu-Hillmer, C, Serrano-Bedia, A M and López-Fernández, M C (2019) Manufacturing and supply chain flexibility: Building an integrative conceptual model through systematic literature review and bibliometric analysis, *Global Journal of Flexible Systems Management*, 20, doi:10.1007/s40171-019-00221-w

Raut, R D, Gotmare, A, Narkhede, B E, Govindarajan, U H and Bokade, S U (2020) Enabling technologies for Industry 4.0 manufacturing and supply chain: Concepts, current status, and adoption challenges, *IEEE Engineering Management Review*, 48(2), 83–102, doi:10.1109/EMR.2020.2987884

Rialti, R, Marzi, G, Ciappei, C and Busso, D (2019) Big data and dynamic capabilities: A bibliometric analysis and systematic literature review, *Management Decision*, 57(8), 2052–68, doi:10.1108/MD-07-2018-0821

Safriyana, S, Marimin, M, Anggraeni, E and Sailah, I (2020) Spatial-intelligent decision support system for sustainable downstream palm oil based agroindustry within the supply chain network: A systematic literature review and future research, *International Journal of Supply Chain Management*, 9(3), 283–307

Sharma, R, Kamble, S S, Gunasekaran, A, Kumar, V and Kumar, A (2020) A systematic literature review on machine learning applications for sustainable agriculture supply chain performance, *Computers and Operations Research*, 119, doi:10.1016/j.cor.2020.104926

Surjandy, S, Meyliana, W, Hidayanto, A N and Prabowo, H (2019) The latest adoption blockchain technology in supply chain management: A systematic literature review, *ICIC Express Letters*, 13(10), 913–20, doi:10.24507/icicel.13.10.913

Toorajipour, R, Sohrabpour, V, Nazarpour, A, Oghazi, P and Fischl, M (2021) Artificial intelligence in supply chain management: A systematic literature review, *Journal of Business Research*, 122, 502–17, doi:10.1016/j.jbusres.2020.09.009

Viet, N Q, Behdani, B and Bloemhof, J (2018) The value of information in supply chain decisions: A review of the literature and research agenda, *Computers and Industrial Engineering*, 120, 68–82, doi:10.1016/j.cie.2018.04.034

Vishnu, C R, Sridharan, R and Kumar, P N R (2019) Supply chain risk management: Models and methods, *International Journal of Management and Decision Making*, 18(1), 31–75, doi:10.1504/ijmdm.2019.096689

INDEX

Bold page numbers indicate tables, *italic* numbers indicate figures